DATE DUE

Your Health
And
The Indoor Environment

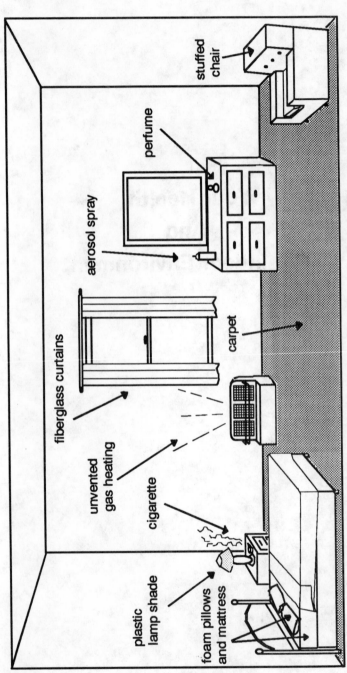

Examples of potentially harmful items in a typical home.

CONTENTS

Appendixes

Your Health
and
the Indoor Environment

A Complete Guide to Better Health Through Control of the Indoor Atmosphere

by Randall Earl Dunford
in collaboration with Kevin G. May, M. D.

NuDawn Publishing • Dallas

Published by:
NuDawn Publishing
10819 Myrtice Dr.
Dallas, Texas 75228

First Printing June, 1991
Second Printing July, 1991
Third Printing September, 1991

Library of Congress Catalog Card Number: 90-92083

ISBN: 0-9628093-3-0

Printed in the United States of America

To Betty and Charles, my mother and father,
who helped me acquire the time I required to
complete this project.

PREFACE

When I first admitted to myself that my health had begun to deteriorate, that those nagging symptoms which had so insidiously crept their way into my life were not going to go away, I hadn't the vaguest suspicion that the most beneficial outcome would be the book you now hold in your hands. Even then, the situation never became obvious to me until I relinquished my job as a route salesman to pursue a full-time career in writing, which kept me home a lot more. At that point, I began to feel worse—still tolerable, but definitely worse—and noticed that my condition seemed to improve somewhat whenever I left the house. But I was a busy bachelor, and even at this juncture I didn't give it much attention.

Eventually, in what seemed like a totally unrelated incident, my sister, Janice Stratton, persuaded me that my bedroom needed a face lift, and she volunteered her services. She went to work, along with her efficient assistant, Yours Truly. We were shocked at what we found in several spots behind the furniture—curling wallpaper, which contained very large, thick patches of mildew. One batch was directly beneath my bed, scarcely a foot from my nostrils as I slept every night.

That day, we managed to remove and discard all the wallpaper, clean the walls, and apply a fresh coat of paint. Still not making the connection, I went to bed that night hoping I wouldn't feel any ill effects from the paint. That would have been all I needed to finish me off.

Fortunately, I was not chemically-sensitive, and not only was I not affected by the paint, I awoke the next morning feeling better than I had in years. It was then that the truth slapped me in the face so firmly, it could have almost created an actual sound. I realized now, that the lethargy, insomnia, shortness of breath, dizziness, irritability, uncontrolable temper (sometimes over even minor inconveniences), constipation, severe sensitivity to cold, severe sensitivity to noise, constantly tense face muscles, and what could only be described as physical depression, symptoms which ordinarily should never have held a rightful place in a man of his mid-30's, had all been brought on by the mold spores which had been unmercifully wafting their way from the mildewed wallpaper into my respiratory tract. Only now did I chide myself for not having paid heed to that "funny odor" I had been noticing every night before retiring.

I rejoiced over my newly-regained health, but it was short-lived. Soon some of the symptoms returned. They were not as pronounced this time, but they were definitely back. Puzzled, I did a considerable amount of probing, and found more mildew hiding behind furniture on the walls in other rooms. I cleaned it up, felt better for a couple of days, then noticed the symptoms returning yet again. I dug deeper and located mildew along the grout between the bathroom tiles, in forgotten corners of closets, and even on the back of book shelves. Again, I rectified the problem, again I felt better, and again within two days the symptoms returned.

As my body became more and more sensitive to the spores, I was forced to continue my "detective work," whereby I tracked down mildew inside my two window air conditioners (so thick in one unit it was almost a solid layer), under the toilet tank, and finally even between the walls of the house. I took care of the air conditioners and the toilet, and sealed cracks in the plaster where I could detect the now all too familiar odor of mildew. This was good for only another two or three days of relief, and by now I was growing both weary and desperate, and seriously considered the prospect of moving. Obviously my home was still contaminated.

Before it was over, I was forced to install central air and heat (which eliminated my need for the unvented gas stoves that had kept the atmosphere humid and encouraged the growth of the mildew), and purchase new carpeting, new upholstery, a water-trap vacuum cleaner, and a high efficiency particulate air filter.

At last reaching the top of the deep dark hole I had begun digging for myself for almost a decade, I realized I had something—something that demanded to be shared with millions of unsuspecting victims who deserve to be informed of the various pollutants which can affect their health and the means by which to deal with them. I was after all a writer, and now I had something more important to write about than I ever had before, the opportunity to publicly point a long and steady finger at these common substances, some of which one might easily fail to recognize as a potentially significant hazard, if indeed he or she suspects at all. I presently set about the task of exploring the subject of indoor air pollution.

It has taken many months (interspersed with what seemed at the time to be more than an unfair share of trying moments), but it has proven to be the most rewarding time of my life. And to think it might never have come about had it not been for Jan, who helped me discover the importance of a clean indoor environment, saving my health and possibly even my very life.

Thanks go also to my primary source of encouragement, Dr. Kevin May, who not only snatched countless moments from an ofttimes hectic schedule to assist me with those pesky fine points that are often so easy to misinterpret, but who also delivered me from the bondage of the dark ages by furnishing the word processor; to Ginger Hardin of Rainbow Vacuum Cleaners, who supplied me with some desperately-needed information; to Jack and Twila Wallace who pointed Ginger in the Dunford direction; and to Posie Hundley of Air Purification Systems, Inc., who played his part in helping me feel well enough to write this book by furnishing a high efficiency particulate air filter. My thanks go also to Jack Harris and J. S. Byrd for their assistance.

In addition, it would be an unforgivable injustice not to express my appreciation for the published works of those physicians who have chosen clinical ecology as their method of practice, as well as for those of the other authors cited within these pages who have helped the world realize the significance of controlling the environment as a means of maintaining better health.

Dallas, Texas   R. D.

INTRODUCTION

For those who never suffer from any chronic symptom, be it a runny nose, constipation, headaches, or even depression; for those who are never plagued with frequent, lengthy infections; or for those currently ill persons who have every confidence that they will make a swift recovery any time now and escape further illness for the rest of their lives, this book is not for them. Don't get things wrong. Never can it be said that every pain, every twitch, every disease is caused by air pollution. But, as will be illustrated in the forthcoming pages, it is becoming increasingly evident that this very problem, especially in regard to the *indoor* environment, is attributable to an alarming number of ailments.

Let's face it. No one wants to be sick. Whether these health problems are a petty annoyance or severely crippling, whether they are termed maladies, ailments, reactions, disorders, diseases, illness, or simply allergies, afflicted individuals want nothing more than to be rid of them. While the information within this volume can offer no guarantee that everyone's health difficulties will be magically erased, it must inevitably prove to be the answer, depending on the nature of the situation, at least in part, for some.

Covered within these pages are the air pollutants common to the indoor environment (with a focus on the home, since that's where we spend most of our time), the symptoms each are most likely to bring about, and the answers for the control or elimination of them.

A word of warning, however. After exposure to the information outlined, it is very possible that you will be flung into the midst of confusion and may find yourself throwing up your hands with the exclamation, "Even if I could find the time to follow every procedure, it would be impossible to do so without sometimes actually trading one environmental hazard for another. It would be easier to just stop breathing."

For instance, how can one open the windows to clear the house of cigarette smoke without inviting in an abundance of pollen grains and dust? Or how can one patch the cracks of a basement to block the flow of radon without chemically contaminating the air with the sealant that is used? Or what good does it do to convert from gas to electric heating if the high operating temperature of the heating elements are going to alter the chemical structure of the dust and render it more harmful?

In response to this, it must be stated that, since obviously, no one is expected to obediently perform every suggestion proposed, there is no intended inference that the only remaining answer is to pitch a tent in the back yard, stop going to the office, and cease buying all available products. Instead, it is the purpose of this book to simply point out every item that poses a potential health hazard, and further, to encourage you, the reader, to make an earnest attempt to determine the source(s) of yours and your family's particular problem(s) and focus on them, while keeping the remainder of the information in the back of your mind.

Even if under the care of a physician, it can do no harm, and quite probably a lot of good, to perform some detective work. Only *you* are capable of recognizing in what part of the house you spend most of your time, what hazards might be awaiting you every day at work, or that every time you are engaged in certain indoor activities, you find yourself feeling under par. Put simply, it is a matter of tracing down the objects, if any, that are most likely to be at the root of the difficulty by experimentation, by using that good old-fashioned process of elimination and separating yourself from them one at a time for a reasonable period to see if improvement results. Habits may have to be changed, certain household objects may have to be discarded, and in some cases, relocation or job switching may be the best answer.

Often the solutions will be obvious. Perhaps discontinuing the use of perfume will reduce or stop incessant rounds of headaches and nausea, maybe eliminating excess dust will put an end to a sinus problem, or possibly frequent asthma attacks will become a thing of the past after a gas stove is removed from the premises. Still, the matter can become more complex if a number of pollutants are interacting in tandem to produce a symptom or a number of symptoms. Moreover, poor nutrition can further complicate matters. In the worst scenario, an individual can react to almost everything, as in the case of Jean Enwright, a Canadian resident who published her account telling of such numerous and debilitating symptoms that she could not even approach a store because of the cleansers, detergents, sprayed produce, or even the chemical additives in clothing, and who was ultimately forced to dispense with all her household cleaners and cosmetics, most of her clothes, furniture, curtains, and linens, her gas heating system, and even her pet cat.[1] More often, however, a few basic changes can work wonders.

Also to be brought to your attention is how this book is organized. First, for those avid readers who faithfully pour over their reading material from cover to cover, a certain degree of redundancy will be discovered, supplied for the sake of those who may choose to home in on only one or several chapters and would otherwise risk missing some useful detail. In addition, most pollutants closely tie in with one another and each are deserving of thorough treatment.

Secondly, some subjects may be categorized differently from the way you might expect. For instance, home furnishings appear in the construction materials chapter, because of their close relationship, including carpeting, which could be considered either. Other heating fuels are covered in the natural gas chapter, since they are all closely related. Although gasoline is a chemical, it is not covered in chemicals but rates a discussion of its own because it is most specifically associated with the automobile, which itself is covered since some of the combustibles from the exhaust often wind up in the house anyway, and since technically speaking, riding in a car is considered being inside. Plastic, too, is nothing more than a group of chemical compounds, but not normally recognized as such, hence it earns separate treatment. And formaldehyde is found in a wide variety of products, but is addressed in detail in construction materials, since that seems to be where it does the most damage to the human system.

Whatever the case, after looking over these pages, you should be well informed about indoor air pollution. Hopefully, this knowledge will lead to an increased quality of life, perhaps at the very extreme, save that life. Good health and a long existence to all.

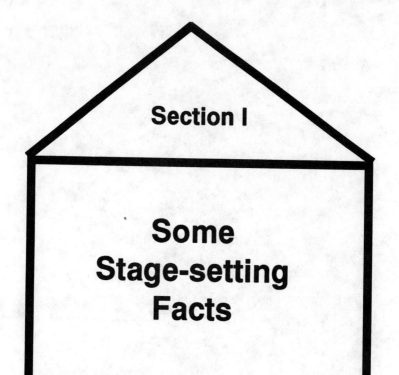

Section I

Some
Stage-setting
Facts

THE BODY'S BATTLE
FOR SURVIVAL

Stop and think for a moment. How many ways are there to get sick? Hint: Even the most thorough person should have no need of a pencil, a long sheet of paper, and a lengthy period to study the matter.

First let it be understood that it would be unfair to include such possibilities as a night out in the cold, an afternoon of dashing through rain puddles, or six months of overwork as causes of illness. These actions in themselves only serve to lower the body's resistance by imposing stress, as does negative emotions (worry, anger, fear, and the like), and the trauma of surgery and physical injury. Although every person's resignation to illness is dependent to some extent upon the degree of that stress as well as their particular genetic make-up, such is not the *cause* of bodily dysfunction.

Avenues Leading to Illness

Actually, the question that should be asked is: How many ways are there for disease-potential substances to gain access to the body? After all, there is virtually no way for illness to occur except by means of these contaminants, whether in the form of bacteria, a toxic chemical, or an abundance of simple unadulterated dust. That narrows the field considerably, making a seemingly complex answer quite simple.

In other words, the list is strikingly short—certainly short enough to comfortably accommodate a single set of fingers.

One of the possibilities is that of absorption through the skin. Unfavorable bodily reactions have been known to occur from clothing, bedding, close contact with chemicals, and even from artificial limb parts. But the human skin is a hearty barrier and this situation is not considered a major factor.

Another thing we must acknowledge is sexual indiscretion. It certainly opens up the opportunity for the introduction of AIDS and many venereal diseases. However, this situation is under the control of the individual and in most instances correct decisions will prove an adequate safeguard.

Of course, one could also receive a harmful substance through injection, whether an illicit drug, an outright poison, or via a contaminated needle. Again, this circumstance is dependent upon correct decisions as well as proper precautions if the administration of medication is necessary.

That leaves us with two remaining concerns—ingestion and inhalation. These are by far the most significant.

No one these days can deny the importance nutrition plays in preventing illness. We make use of our digestive system 3 times a day (more, if we conduct a midnight raid on the fridge). This presents plenty of opportunity for the less discerning of us to keep our digestive systems clogged with any number of unwholesome foods. But publications on the subject abound. TV and radio ads continually tout all-natural products. There is no one who has not become well acquainted with terms such as "processed foods," "preservatives," and "artificial sweeteners," and the fact that many additives have been discovered to be carcinogenic. The idea of balanced meals for proper vitamin and mineral intake and a minimum of fat is repeatedly advocated. We have become a diet-conscious society.

That leaves us with the issue of air quality. It is no less important in regard to health, and perhaps in a way it is even more so because to survive, nature demands that we breathe constantly. That makes exposure to any present airborne substance(s) perpetual. Furthermore, due to the smallness of its some 300 million alveoli (air sacs), the approximate internal surface area of the lung is an astounding 750 square feet (the size of the floor plan of a small house)—much too much to be open to the myriad of modern-day pollutants.

These particulate and gaseous contaminants have increased dramatically during the last half of this century. Pesticides, gasoline vapors, and cleaning chemicals are just a few that have contributed to the cause, not to mention the age-old enemies such as dust, mold, bacteria, and pollen. It seems the adage "You are what you eat" can in this case be justifiably altered to "You are what you breathe."

There are acute cases of air-pollution-related illness, of course, but they are more often attributable to obvious conditions such as severe episodes of smog, transportation accidents involving hazardous materials, or incidents such as the one in Bhopal, India in 1984 or the one at Chernobyl. Much more common, but all too often discounted, are the smaller, yet still potentially-potent doses of airborne matter that are gradually, insidiously sneaking their way into our body with every passing minute, creating those annoying and sometimes alarming chronic disorders.

In fact, according to the 1979 U. S. Surgeon General Report on Health Promotion and Disease Prevention, there is virtually no major chronic disease that is not either directly or indirectly produced at least in part from poor environmental conditions. And if that isn't enough, Dr. Kevin May, an M. D. practicing internal medicine in Tyler, Texas, reminds us that chronic ailments often "open us up" to the possibility of acute illness.

The Body's Limits

Even undergoing the daily abuse that air pollution dishes out, however, the human body proves itself to be a remarkable piece of equipment. In a complex, yet efficient manner, it possesses the ability to perform feats such as those of assimilating fuel (food), eliminating wastes, repairing minor wounds inflicted upon it, serving as short-range transportation, coming up with some of the most dandy ideas, accomplishing any reasonable task, (and all too frequently, some that are not so reasonable). In short, it is a walking, fleshly miracle—the most precious commodity in existence.

And not only does it perform every demand placed upon it under ordinary circumstances, it possesses unbelievable tolerance, faithfully fighting the silent invading armies of wanton particles. The nasal hairs screen some of these offending substances—those over approxmately 5 microns. (A micron is a linear unit of measurement equal to 1 millionth of a meter, or 0.000039 of an inch. For purposes of comparison, the diameter of a human hair is 60 to 80 microns.) The remainder, still a whopping 90%, must be dealt with by the cilia (tiny fibers) of the lungs and trachea (windpipe). These cilia, along with mucus that has been created for this occasion, push the contaminants back through the upper respiratory system in a somewhat "escalator effect," where they can be excreted by coughing, sneezing, or blowing the nose.

But the pollutants that the cilia can't handle remain in the body where they are, to put it simply, fought by cells that absorb and destroy them or produce antibodies to nullify their effect. The human system, in essence the site of a microscopic battlefield, relentlessly expends a great deal of energy to manufacture this protection whether it is continually inundated with cigarette smoke at the office or airborne mold spores in the bathroom.

However, even a loyal, hard-working device such as this has its limits. Like the deteriorating automobile in which the engine's crankcase has become increasingly dirty due to neglect, its defenses wear thin, and tiring of the grueling overwork, it finally lets its troubles be known.

The countless victims that have suffered with far more than their share of symptoms can attest to this. Often, allergenic responses develop such as coughing and sneezing. Sometimes excessive amounts of mucus become infected, which can inflame the bronchial tubes, leading to problems such as chronic bronchitis. On other occasions, as the airways become blocked and it becomes harder to expel air, pressure is created on the delicate alveoli. If this condition gets bad enough, the alveoli can tear, and this causes what is known as emphysema. In addition, moist air can stagnate in the lungs because of this condition, and becomes an ideal site for bacterial growth. Of course, other respiratory disorders such as sinusitis, asthma, and lung cancer also, all too often, make themselves known because of tainted air.

It's no wonder. The daily dose of pollutants that the body must endure is frighteningly large. Of the 21,000 to 58,000 breaths we take each day (the exact number, of course, being dependent upon our rate of activity), we inhale several thousand gallons of air, and Sharon Faelten, author of *The Allergy Self-Help Book* tells us that within that volume there exists at least 2 heaping tablespoons full of assorted contaminants. That's far too much to be ignored.

Kinds of Disorders Caused by Air Pollution

However, one might venture, "My family and I are not menaced by any sort of breathing difficulties. Is that not substantially the kind of thing that results from air pollution?"

It's true not everyone suffers from respiratory dysfunction. But unfortunately, the physical problems of air pollution extend a lot farther than that.

To understand this, let's review some basic biology. When we inhale, obviously oxygen is drawn into our lungs. From there, it is absorbed by the blood-vessel-rich alveoli walls (which, by the way, are a mere one-millionth of an inch thick), diffuses into our bloodstream and, in turn, is carried by the red blood cells to every part of the body. Therefore, most anything else in the immediate atmosphere which is taken in during the breathing process—and escapes the cilia—is likewise distributed and must be reckoned with by all the human organs if the body's other defenses are no longer in control of the situation.

The unwanted culprits can, in addition, enter the digestive system from swallowed particle-laden sputum that one may fail to eliminate when coughing or sneezing. Pulmonary infections such as tuberculosis have been known to be diagnosed because of temporary flora in the stomach. And if that isn't enough, they can be absorbed back into the blood from there.

There have been many documented cases of non-respiratory ailments that have been linked to various forms of air pollution. A noteworthy one involved a patient of Dr. Alfred Zamm, a Kingston, New York physician who practices clinical ecology (a branch of medical science that focuses on curing disease by improving the environment). In this instance, a clothing-store owner consulted him about a severe hand rash. Dr. Zamm recommended that he discontinue the use of floor wax and mothproofing in his home and store. As an added precaution, the patient also dispensed with his wool living room rug and wool blankets (since he was also found to be somewhat allergic to wool), and for good measure even employed a dustproof cover for his mattress and installed an air filtering system. Not only did he find relief from his original problem, but he was surprised to note the sudden, mysterious absence of severe headaches which he had long since taken for granted because he had suffered with them all his life.

Another example comes from Dr. Marshall Mandell of Norwalk, Connecticut, also a clinical ecologist, who has conducted a number of test studies which have aided him in diagnosing environmentally-produced illness. He discovered that a female patient's sudden tension and abdominal cramps, not to mention her stuffy nose and itchy eyes, were caused by the gas range she cooked on. Still another example is related by Elaine Bonavita Jaquith who, in her book *Allergic to the Twentieth Century*, describes how, among other bodily dysfunctions, her nausea and frequent sleepiness were caused by pollutants such as perfume and gasoline vapors.

And the list goes on.

Okay, now that there seems to be no doubt that the full spectrum of physical illness is possible from a contaminated atmosphere, let's move a step farther. Is it possible for air pollution to cause mental disorders?

According to Dr. Mandell, "In most instances, mental disorders are reversible malfunctions of the brain that are caused by the afflicted individual's susceptibility (sensitivity, reactivity) to many commonly encountered substances."

To clarify this, we must move again to the bloodstream.

Remember, every body part is dependent upon the oxygen carried to it—including the brain. In fact, it has long been established that the brain is the most quickly damaged organ when deprived of oxygen. Is it any less wonder that it can be impaired by impurities in the blood?

This opens the door to many a case of unexplained behavior, which, unpleasant enough for the victim, frustrates loved ones, strains family relationships, and has most assuredly even been responsible for incidents of divorce. Hyperactivity, belligerence, temper outbursts, social withdrawal, personality changes, and even the more extreme occurrences such as nervous breakdowns and criminal conduct have sometimes been linked in full or in part to contaminated air.

Dr. Doris Rapp, a Buffalo, New York, pediatric allergist, tells of a teenage boy who was taken to a psychiatrist because of constant horrendous behavior. But the strategy proved fruitless and his rebellious habits continued until his mother finally discovered the secret. Although he also had a number of physical and mental symptoms brought on by certain foods, mold spores and chemical odors had been playing a significant role in his behavior problems.

An even more extreme example is a published case of Dr. Mandell's, involving a schizophrenic woman who had been to five mental hospitals. He discovered that, along with several foods, cigarette smoke and insecticides were responsible for the condition.

In another reported case, chemicals commonly found in office buildings proved to be the villain for a man who was plagued with cerebral symptoms so severe, he couldn't even appear in person to apply for social security benefits.

It is evident, then, that being equally exposed to foreign airborne matter, no body part—whether a nerve, the skin, or a deeply-planted organ—is exempt from the possibility of malfunction. Clearly, any ailment (physical or mental, pesky or painful, allergic or catastrophic) can develop, depending, of course, upon the nature of the substance(s) absorbed and the degree and length of exposure. It has even been known for a toothache to result from harmful inhalants. In the last 2 decades, the untold thousands of documented cases of air-pollution-induced ailments from watery eyes to hallucinations to kidney problems cannot be shunned by the most devout skeptic.

As Dr. Mandell sums it, "Reactions to commonly encountered and generally unsuspected environmental agents frequently cause a most amazing variety of reversible physical and mental disturbances involving the field of psychiatry and all of the medical and surgical specialties."

It's a shame the issue of air quality has not received more recognition than it has. Chronic body ailments have become more the rule than the exception. According to the National Center of Health Statistics, over 31 million people suffer from sinusitis alone—more than any other chronic medical condition. A close second is arthritis, followed by high blood pressure. And that's not to mention other common concerns such as headaches, abdominal pains, constipation, nausea, fatigue, irritability, insomnia, and memory loss to name just a few more.

There's no telling how many of these cases involve poor air quality. Come on, admit it. Even you regularly experience some brand of bodily discomfort. Virtually no one can boast that they have gone untouched. In fact, Dr. Zamm ventures that "everyone is environmentally susceptible to some degree."

The Sense of Smell

Fortunately, though, there is a silver lining. The body possesses a wonderful warning system that can help greatly in avoiding many of these problems. It's called the olfactory sense, or sense of smell, unfortunately the one that seems to have been given the least amount of attention. But it is no less deserving. Estimated to be some 10,000 times more sensitive than the sense of taste, it is capable of perceiving from 2 to 4 thousand different odors. Even extraordinarily low concentrations of molecules from odor-producing substances can be detected.

One tends to take this valuable ability for granted unless entering a chemical factory or a garbage dump. But if one develops it, uses it to advantage, he can detect some surprising things during his daily routine—things that he might not be intended to habitually breathe.

Remember the smells encountered when at an amusement park? How about a bowling alley, a movie theater, or a clothing store? Each have their distinctive aromas—or odors, depending upon the substances within, or in some cases, the individual's opinion. But we are so inured to them, there in their "rightful" places, that we usually accept them with little or no thought. As long as one sniffs the evidence of cooking popcorn at the theater or of the emanations of chemicals from the treated garments in the clothing store, everything is in its place and all is well. Yet the presence of bowling alley odors in the theater might just elicit some sudden attention.

The dwellings in which we reside present no exception to this rule. Virtually every home harbors one or more perpetual odors. That reality was dubbed "housitosis" in a TV commercial a few years back, and it should have rated a position in the dictionary. We have become so familiar with it, it has become such a part of our lives, that unless it has developed into an extreme case, very few of us have probably ever even given it more than the most casual of thoughts.

And to compound the problem, shortly after one enters a room, building, or any other odorous area, even if initially aware of the existing smell(s), he experiences what is known as olfactory adaptation. That is, he ceases to have the ability to detect the odor—regardless of how strong it might be.

This can prove dangerous if one fails to notice and heed the initial warning, as in the loudly publicized example in Dallas a number of years ago. It seems an entire family was confined to the house with what they thought was the flu. The condition had hung on for an extraordinarily long time and they wondered why none of them could shake it. Fortunately, they were eventually visited by a friend, who, upon entering the premises, quickly gave the command to get out of the house. "What?" they asked, "March our ailing bodies out into the winter air? Preposterous!" Then the friend explained he was being assailed with such a strong odor of natural gas, he thought the place was going to blow any minute. A leak was subsequently discovered, and needless to say, all the members made a rapid recovery.

It is far too easy for one to continually expose oneself to such airborne hazards without ever realizing it, whether it be within the home, office, or some other frequently visited place. That is not to brand any particular substance as necessarily harmful in every instance, but all too often this is indeed the case. Scores have suffered because their bodies have constantly been inundated with something in the air that really shouldn't be there.

Try the "sniff test" the next time you go back into the house. Stale cooking odors, new carpet smells, or vapors from chemical cleaners may be permeating the air. And such things just might be affecting you. Bear in mind, too, that more than one contaminant could be involved, and that the odor of some equally threatening ones could be masked by those of a more penetrating nature.

Also, be no less aware that just because an odor is pleasant does not mean it cannot eventually cause harm. Although many are often repelled by smells that give them problems, some can become addicted to certain potentially-harmful odors and thereby make no effort to avoid them.[1] Equally important, just because your "sniffer" fails to detect an odor does not mean there is nothing hazardous in the air. There are odorless substances too—some deadly, e. g. carbon monoxide.

Importance of Being Aware

Sadly, the countless number of individuals walking this planet today who are plagued with symptoms brought on by impure air are not the wiser as to the reason why. Many may not even realize they have developed a sensitivity because the resulting symptoms are, at present, so mild or because they are simply too busy to notice. They may be fretful day in and day out for no apparent reason, never realizing that the paints and glues they have been employing in their model-building hobby is the answer. Or maybe the joint pain and the depression (two seemingly unrelated symptoms) that beset a 50-year-old housewife is accepted because she thinks, "I'm just getting old," when the reason may lie with the chemical cleaner she regularly uses to scrub the kitchen. Or perhaps a young man's headaches have come on so gradually because of cigarette smoke that he just doesn't as yet realize how frequent and painful they have become.

Fortunately, there are victims who are becoming aware of this plight. As they do, they obviously want nothing more than to dispense with the thing(s) that is causing their misery—a worthy desire, because the longer the body is exposed to the substance(s) it has become sensitive to, the more that sensitivity is likely to increase, and certainly the more strain there is on the body.

A good phrase to remember is, "more strain, more pain," because as the body weakens, like a worn filter it progressively becomes less efficient in protecting itself and may develop not only more susceptibility to what was originally troubling it, but often bow more easily to other pollutants as well. In some cases, one health-threatening inhalant might be tolerable, but the combination of two or more may create definite problems. For instance, one may still be able to get away with breathing an inordinate amount of cleaning chemical vapors, but when that atmosphere is also flooded with a supply of cigarette smoke, the story may change.[2]

It is a pity that so many have been forced to resort to the vast number of medications on the market (over $40 billion spent annually in the U. S.), whose only value in the case of chronic disorders is to relieve the specified symptoms. That might make the person feel better temporarily, but it does nothing to stop the cause. By all rights, such treatment should be utilized only when absolutely necessary, as in cases of infections or emergencies.

If airborne contaminants are found to be the problem (or a part of the problem), the source must be located and removed. Or the person must be removed from the source, whichever is more practical.

There are numerous ways to do so, some simple, others more elaborate if it becomes necessary to go that far. Future chapters (in Section II) will identify the possible causes of one's ailments and describe the actions that can be taken to eliminate or control them.

PAINTING THE POLLUTION PICTURE

Now that we have a somewhat better understanding of the workings of the human body, let's clear the air about atmospheric pollution. There are a number of different categories: Light, noise, even electromagnetic radiation qualify as pollutants. But the one of extraneous gases and particles is the most widely recognized—and rightfully so, because it plays the most important role in the status of human health.

When we think of the term "air pollution," we automatically associate it with smog, smoke stacks, auto exhaust, and in short, any undesirable airborne substance connected with the out-of-doors. It is true, of course, that a great amount of it does originate from there, giving rise to a mounting problem. It is a well-established fact that modern society has fallen prey to its industrial and technological progress. Five hundred million pounds of chemicals are produced annually. The number of compounds in common use in this country today is estimated at over 60,000—and that figure is growing steadily, adding more to the already ailing environment. It is said that on the average, one new compound is synthesized every minute![1]

But just the items that have been of the most immediate concern are bad enough. The subject of lead accumulation, for instance, has been pushed to the forefront in recent years. Although strides have been made to reduce the hazard, Frank Mitchell, a specialist with the Atlanta-based National Center for Disease Control admits that lead poisoning continues to be a problem with children in this country. And that's only one example.

Regulators have been forced to demand stricter compliance regarding environmental control at coal-fired electric power plants because of acid rain. Much talk has centered around the ultimate replacement of the combustion engine with an electric variety. Even asbestos dust is a concern because fibers of this material are regularly ground off of brake linings and clutch plates.

And the list goes on.

Pollution and the Indoor Environment

But that's not the whole picture. Air pollution is even more far-reaching than many realize. We certainly don't see signs adorning the entrances of homes, schools, and businesses containing the message, "Attention outside contaminants: admittance beyond this point forbidden." No matter how demanding, pleading, or eloquent, words will exert no influence in this direction. Neither will the performance of some flashy ceremony, nor an act of magic. The harsh truth is that all airborne substances which originate out-of-doors are capable of finding their way inside.

They sneak in on shoes, clothes (We all know what a dirty dryer filter looks like.), hair, pets, insects, and as part of the air itself. Depending on the nature of the substances, and general conditions such as temperature, humidity, and the amount of foot traffic, they can settle on carpeting, upholstery, or other objects, only to be subsequently stirred and recirculated from the slightest human activity. They can even land on exposed food and later be ingested. Or they might remain aloft, riding the tenuous currents, lingering for that possible journey into the lung.

The quantities are minute, certainly—but cumulative. The theme of the 70's and 80's has been one of increased energy efficiency, and so with the addition of insulation, storm windows, caulking, and the like, most structures, both public and private, now have far less ventilation. Older houses, unless augmented with energy-saving features, have an air-replacement rate of once every hour or so. But it requires several hours—sometimes as many as ten—to turn over the same volume of air in other houses. Impurities have much less opportunity to exit. What's worse, typical heating and air conditioning filters can be expected to trap no more than 10% of this harmful matter. Besides that, much of it is surprisingly resistant to conventional cleaning methods. (See chapter 5, *The Dust Dilemma*.)

Also, in addition to what outside contaminants contribute, it is alarming the amount of pollution we manufacture right inside our places of residence and employment. Hopelessly trapped are cigarette smoke, cooking residue, vapors from perfume and cleaning chemicals, unspent heating fuels, bits of sawdust, lint fragments, and a host of other microscopic junk.

Numerous studies have been done on the subject of indoor air pollution, and they speak for themselves. Here are just a few examples:

■ Experiments conducted by Yale University in 1976 revealed that the air in 19 out of 20 homes tested was considerably more polluted that the outside air.

■ A research team from Lawrence Berkeley Laboratory analyzed the air in six typical American (San Francisco) homes and found concentrations of noxious vapors such as carbon monoxide, nitrogen dioxide, and nitric oxide which was 4 times the maximum amount recommended by federal standards.

■ In 1984, a study by the U. S. Consumer Product Safety Commission revealed the presence of as many as 150 different chemicals indoors—compared to 10 or less outside.

■ An Oak Ridge, Tennessee national lab study uncovered levels of volatile chemicals in the air in 40 homes of various ages that were 10 times the outside level.

■ In 1985, the EPA went a step farther by announcing that a completed study revealed concentrations of dangerous pollutants in indoor air which was as much as 70 times that of the outside atmosphere!

But here's the clincher. According to the American Lung Association, the average person spends at least 90% of the time indoors (65% at home). When not at home, that person is likely to be at the office, a restaurant, a store, someone else's house, or in transit to one of those places. That's not too far short of maximum exposure! Sadly, anyone who tries to avoid an outside contaminant such as gasoline vapors or pollen by

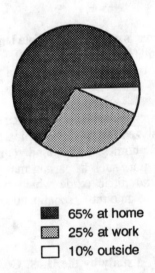

■ 65% at home

▨ 25% at work

☐ 10% outside

Amount of time spent indoors. The significance of maintaining a clean indoor environment is obvious. The average person spends approximately 90% of his/her time inside. Even those employed for outside jobs may often be required to spend some time indoors as well as time inside a motor vehicle.

remaining inside even more, may very well not only be failing to escape its grip, but actually be adding to his woes.

This would suggest such a situation to be especially true in winter when one's period of confinement is at its peak. Is it any coincidence that many acute respiratory ailments are so much more prevalent that time of year than in warmer weather when the outside air is capable of holding a far greater quantity of matter anyway?

Indoor air pollution has been labeled "an emerging health problem" by the comptroller general of the United States. Moreover, Dr. Theron Randolph, a renown Chicago allergist and clinical ecologist, tells us that in his experience, it is "eight to ten times more important as a source of chronic illness in susceptible people than ambient air pollution."

Environment: Outdoor vs. Indoor

In fact, although dreadfully taxed by extraneous matter, the outside atmosphere is periodically cleansed by nature's built-in system. Winds sooner or later move out excessive accumulations in major industrial areas. Eventually, rain and dew send particles earthward where many combine with the soil. Likewise, cooler temperatures force the air to shed much of its unnecessary content. At one time or another, all have detected the freshness of an autumn breeze or the clean, sweet scent that follows a healthy rain—even in the most bustling city. Additionally, of the some 1 quadrillion (1 followed by 15 zeros) tons of oxygen present, it is estimated that 400 to 500 billion tons are produced yearly by means of photosynthesis. Now that's a whole lot of diluent.

So there is ample room for optimism in that no one need be in constant fear of the out-of-doors because of air pollution. At least the atmosphere gets properly recycled. That beats the indoor environment, a stale and toxic prison, reputed by modern man to be an effective shelter from the synthetic, as well as natural elements.

It's a shame the number that have found themselves mired in the throes of an illness which they never suspected could have been aggravated or created by the extraneous content of the air within the walls that surround them. Probably the most vulnerable in this respect are the elderly. Not only has their exposure time been increased due to their larger number of years, most have retired, slowed down, and kept themselves closed up in their houses or apartments all the more. Often their shades remain down, perhaps because of their sensitivity to the light, another reported symptom of environmental illness. Shut off from the sunlight that interacts with the skin to activate vitamin D (which in turn aids calcium absorption), and possibly neglecting diet, the problem is compounded. Their senses fail. Their energy is lessened (yet another symptom often created by poor air quality), which forces them to forsake cleaning chores, lending to the excess accumulations of dust and other particle matter. Perhaps in their cosiest corner is an abundance of mildew, or animal dander from a pet. Or maybe they fail to detect the persistent vapors from spilled kerosene in the basement.

Infants and invalids are also especially susceptible, as is the expectant mother. Not only are they in an indisposed condition, they spend more time at home because of it, which puts them all the more at the mercy of their immediate surroundings. Even older children are sensitive to the effects of air pollution due to their greater rate of breathing and smaller lung size.

Of course, no member of society is exempt. A noteworthy example of inside health problems fell within the experience of Dr. Randolph. He recommended the removal of a gas range from the home of one of his patients, a girl suffering from persistent headaches. Not only did the move benefit the girl, but her mother unexpectedly profited; she soon realized that she was no longer getting irritable every time she cooked.

Whether we wish to face it or not, our indoor environment is continually being bathed in countless noxious substances which is nothing less than a grave threat to our health. According to the National Academy of Sciences, indoor pollution may contribute as much as $100 billion to national health care costs. It is imperative that we become attentive to these seemingly innocuous concentrations and determine if any of them are spelling trouble for us.

That's the general background. Now let's get down to some specifics.

Section II

The Culprits: Identification, Health Effects, and Solutions

3

THE UNNATURAL ASPECTS OF NATURAL GAS

Natural gas is largely composed of methane, a colorless, odorless, flammable substance which is formed by the decomposition of vegetable matter. There was a time when oil drillers would allow any gas they encountered to burn at the well because there was no way to transport it over long distances. But soon man began to establish a pipeline network, thus a new industry was born.

Now widely employed throughout the world as an alternative to burning coal, oil, or wood, it has been hailed as a clean fuel because it does not produce any kind of visible residue. Nonetheless, nothing need be seen to be harmful, and in fact, natural gas does create its own brand of pollution.

Components of Natural Gas

For starters, the major constituents of natural gas are hydrocarbons, which are released during incomplete combustion. (No heating device, no matter how perfectly designed or maintained, is 100% efficient in burning fuel.) Furthermore, there are chemical additives present, including the one that gives the gas its familiar odor to aid in the detection of its presence. (The consistency of that odor, by the way, can vary in different parts of the country because different chemicals are used.) In other words, every time a gas heater or oven is turned on, the burner may be pulling into the home substances such as ethane, benzene, acetylene, formaldehyde, carbon monoxide, nitric oxide, and nitrogen dioxide as well as hydrogen sulfide (one of the odorants), which not only contributes to pollution itself, often resulting in the irritation of eyes and respiratory tissue, but eventually causes pipes to corrode, initiating gas leaks.

In addition, natural gas has been known to be contaminated with polychlorinated biphenyls (PCBs). An investigation in 1981 by OSHA (Occupational Safety and Health Administration) turned up 150 parts per million of PCBs in some of San Francisco's utility lines—three times the EPA's hazard level.[1] Similar situations have been reported in Long Island, New York and elsewhere.

And that's not all. Believe it or not, it is also possible for natural gas to contain radon because sometimes both exist in the same geological strata.

Health Effects

There is no doubt that this "natural" gas looms as one of the greatest contributors to indoor pollution and therefore human illness.

In one study conducted by Dr. Mandell, natural gas was responsible for unfavorable reactions in over one third of 30 patients tested.

Dr. Randolph goes on to state that "natural gas is highly *unnatural* as far as the human body is concerned—a substance with which the body has no physiological method of coping."

The danger brought about by major leaks is, of course, obvious. Over a thousand people are thought to lose their lives annually because of gas poisoning in the home. But just the amount that is liberated whenever one, say, hits the valve control to light a space heater has been known to affect particularly sensitive individuals. For that matter, a pilot light can, over time, do the same thing. If that's not enough, even when appliances are shut down, minute amounts of raw fuel (some too small for even the gas company to detect) invariably escape from pipe joints and the valves themselves. After all, the integrity of every physical structure is less than perfect. How many times have you picked up a whiff of gas when passing a meter outside?

Still, the story is not complete. People have been known to be bothered by disconnected gas appliances. Equipment such as ranges and heating stoves can be totally removed from the utility line and all the same perpetuate an illness because the metal has the ability to absorb the vapors.[2] The older the appliance, the worse this condition will be.

To get a grip on just how serious the overall situation is, several studies have been conducted and the results were appalling. It was discovered that a typical gas oven operating at 350° Fahrenheit for one hour in a vented kitchen customarily dispenses a level of carbon monoxide and nitrogen dioxide equal to the level reached in Los Angeles during periods of heavy smog.[3] Even one operating burner can produce more than half that much. What's more, if the room contains no means of exhaust, the condition can become as much as 3 times worse. In fact, under this circumstance, the carbon monoxide content alone can exceed 100 ppm (parts per million). (A figure of 35 ppm represents the level of allowable exposure according to U. S. Government standards.)

Unvented gas heat can prove to be especially bad news not only in that it allows a more potent build-up of vapors, but because it gives rise to excessive moisture in winter, which in turn encourages the growth of mold. (Consumers, incidentally, stand to save in the neighborhood of 15% on fuel bills by venting their heating systems.) Even a newly-installed gas-fired central air system, vented as it may be, can contribute nitrogen dioxide and nitric oxide at levels 3 times that of those recommended by federal air-quality standards.

Unquestionably, there are many smitten with winter ailments—some in a constant state of illness—that could very well be the result of perpetual exposure to natural gas, even if only in trace amounts. The most likely resulting symptoms are the same ones associated with any chemical sensitivity, mainly the physical problems of headaches, fatigue, nausea, and dizziness; and such mental concerns as depression, irritability, and confusion. It is not even unreasonable to believe that SAD (Seasonal Affective Disorder), known to many as the "winter blahs," can be brought on by the use of natural gas.

The situation can sometimes be worse than that, too. The content of nitrogen dioxide in natural gas, over time, is capable of damaging lung tissue. It can also lower one's resistance to disease by affecting the white blood cells. Several British studies have revealed a higher incidence of both respiratory symptoms and disease in children 5 to 11 years of age residing in homes with gas ranges as opposed to those in homes with electric cooking. Other studies involving that age group under the same conditions also revealed a higher frequency of coughs and colds.

One rather drastic example of illness induced by natural gas is related by Dr. Mandell. It concerns a young man, an excellent swimmer and Olympic hopeful, whose health inexplicably began to deteriorate. He was beset by energy loss, aches, irritability, depression, exhaustion, dizziness, and light-headedness. His condition became especially bad every time he entered the kitchen, where he would also undergo a personality change and suffer attacks of bronchial asthma. Years of exposure to the natural gas that had been repeatedly pumped into his family's home, particularly from an extra large gas range, had been responsible.

Living With Natural Gas

If you suspect natural gas to be the root of your problems, it might be wise to ask yourself some questions. Do any ailments I have seem to decrease in intensity or disappear in summer? Does my condition improve when I am outside? Am I worse, or do I develop symptoms every time I or my spouse uses the gas range? If there are some affirmative answers here, natural gas could be partially or completely accountable.

However, take heart. The situation is far from hopeless. There are a number of maneuvers that can be instituted to protect oneself from the hazards.

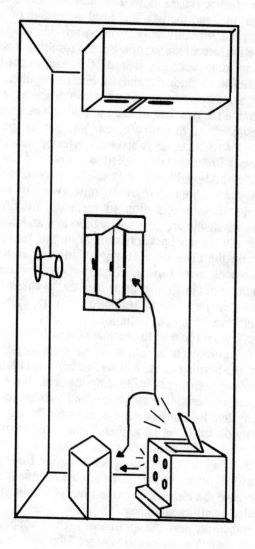

Controlling natural gas emissions. Opening windows and the use of exhaust fans reduce the inside level of pollution when cooking with natural gas.

Gas Ranges

First, as far as cooking is concerned, make sure to install an exhaust fan above the range if the kitchen is unvented. (This is required for electric ranges as well because smoke, moisture, vaporized oils, and the like also need an avenue of escape.) As it is, it's virtually impossible to provide enough ventilation to eliminate all the vapors. But the resulting lower levels of gas, in some cases as much as 50% less, might be enough for relief in individuals who are less susceptible. If nothing better, at least open windows whenever practical, and position a portable fan near one of them to function as a temporary exhaust system. (Take care, though, not to get a running fan to close to a gas range and inadvertently blow out one of the pilots.) Some kind of ventilation is a must if one is to keep the toxic build-up diluted. In fact, it is a good idea for people in tightly sealed homes to partially raise a window even with an exhaust fan in operation to maintain air flow. Otherwise, the inside air pressure can get too low and cause problems such as chimney backdrafts (more on this later).

To minimize the dependency for ventilation, it is suggested that one cook more slowly. The reduced temperature will lessen the rate of contamination as well as the chance of boil-overs which can deposit food residue on burners, thereby creating an additional brand of pollution.

Moreover, it would be wise to consider ranges that employ auto (electronic) ignition rather than active gas-burning pilots. Then, at least, the burning of the fuel will be confined to designated times—not perpetually infiltrating the area. If this is impractical, you can establish an improvised version of an auto ignition system by shutting off the pilots (or having the utility company do it for you) and then lighting the burners with a match whenever the range is used.

Another alternative to controlling air pollution from gas ranges is the use of ceramic burner inserts. It has been said that they help decrease the nitrogen dioxide level. Consult your appliance dealer or utility company for details.

Best yet, substitute portable appliances such as hot plates and steamers or switch to an electric range. That will totally eliminate gas emissions in regard to cooking. Dr. Randolph tells us that in the course of his practice, he has instructed some 3,000 patients to remove their gas kitchen ranges because he found them to be susceptible to the fumes. He adds, "To date, none of these patients has complained that the changeover was not worth the cost or trouble."

Gas Heating

Of course, it wouldn't be so bad if cooking was the only consideration. But since millions of homes (and other buildings) use natural gas for heating purposes, the inside air is continually being permeated with it. Unfortunately, that's too much for many.

Unvented portable stoves are the worst offenders and should be dispensed with at the first opportunity. Building codes now forbid their installation. The electric variety is a much better choice. A heat pump system, an electric baseboard arrangement, or electrically-produced steam heat in the form of wall radiators are also options worth considering, as is supplementary solar heat. As far as gas-fired central air is concerned, the best one can do is keep the thermostat on the lowest practical setting. If one plans on installing a central unit, however, it would be wise to play it safe and get an electric furnace if there is any doubt about sensitivity to natural gas. Anyone who insists on installing the gas version, though, should arrange to place it as far away from the living quarters as possible. An attic would certainly beat the closet near a bedroom. That may make it somewhat less accessible for servicing, but the pilot and burner will not be as close to you—and your lungs.

Don't let it shake you either, but if you are a particularly sensitive individual in this regard, it may become necessary to rid your home of gas completely (including the gas lines).[4] This will entail effort and expense, but if you are desperate enough, you will find yourself willing to do anything. If such a thing does become necessary, don't forget about the water heater, clothes dryer, and in some cases even the refrigerator and/or air conditioner.

If renting a house or apartment and the choice is not yours, it would be advisable to find another home. This is drastic action, granted, but isn't a move better than the constant suffering from some chronic ailment or perhaps the prospect of facing the eventual onset of a catastrophic disease?

A Word About Maintenance

At any rate, if you decide to retain your gas appliances, keep them well maintained. At best, their burning efficiency is only about 80%, and with neglected equipment that figure can plummet to as low as 40%. Make it a habit to have burners checked annually, and upgrade any equipment that is wearing out. Always remember that a gas range burner flame should be blue in color. If the tip of the flame is yellow or orange, this indicates an incomplete burning of fuel and an adjustment is required.

Don't forget that well-maintained furnaces keep heating bills at their minimum. Proper air flow is essential. A heating system can be tested in this respect by holding a hand over the air diverter while the burner is on. You will feel suction if the system is functioning properly. If you detect warm air, it is an indication the flue is blocked. Obviously, the problem will require immediate attention. If you are not inclined to investigate this, check with the personnel at your local gas company. They, of all people, understand the urgency of the situation and will usually examine furnaces if requested.

Other Heating Fuels

Kerosene

While on the subject, its only fair the give attention to the other heating fuels as well. Kerosene (known to some as coal oil) can also be a serious offender. Its vapors, raw and in the burning state, can be just as harmful. If any of the oil is spilled, fumes can remain for worrisomely long periods—months or sometimes years.[5] Even the most ardent cleaning efforts are not completely effective. Many times, oil tanks (especially old ones) will leak and, if they happen to be located in a basement, a constant level of pollution will be wafting its way up into the rest of the house. In fact, a characteristic odor prevails even when kerosene furnaces or space heaters are shut off, indicating some quantity of matter is still riding the air.

Anyone who insists on using a kerosene heater should heed the advice of the American Lung Association. Keep it cleaned and properly adjusted—and never use one that is un-vented. (The sale of unvented kerosene heaters have been banned in some states.) Moreover, it should be filled out-of-doors with a low-sulfur fuel and never used in a closed room or be allowed to burn overnight.

Wood

Burning wood can be a mistake, too. The resulting smoke is harmful enough (rendering such contaminants as carbon monoxide, methane, methanol [wood alcohol], tar, sulfur dioxide, indeno pyrene, fluorene, and nitrogen oxide), but if the wood has been chemically-treated, all the worse. It wouldn't be so bad if all the combustible remains obediently traveled up and out of the house. But it's not that simple. Some amount of residue will always escape into the room. On many occasions, this fact becomes obvious when a backdraft (a big puff of smoke) is released, as is typical with fireplaces. Even when they are not in use at the moment, wind can sometimes blow out soot—not a good thing either.

This problem, especially if persistent, can often be traced to a faulty damper or a clogged flue. Nearby objects taller than the chimney can also direct currents down into the fireplace.

If using a fireplace, make sure the damper is functioning properly and keep it closed when the fireplace is not in use. However, never close it until the ashes are cool and always open it *before* lighting a fire. Have the chimney as well as the fireplace cleaned at the end of every season. It would also be a good idea to install glass doors across the front to confine the soot. If still plagued with excessive backdrafts (also known as downdrafts), it may be necessary to use a chimney cover or cap (often made of brick or a concrete slab built over the top of the chimney complete with openings in the side just below it). Check "Chimney Repairers" in the yellow pages for details.

Incidentally, the use of old newspapers as a fuel, in a fire-place or woodstove, is out of the question. This can contribute lead to the atmosphere, especially if they contain colored ink. Colored ink can also release traces of arsenic vapor and other chemicals.

Coal

Coal, of course, gives rise to coal dust, another health hazard. Even the utmost in handling will result in some measure of airborne particles, not to mention the fact that it, just as any fuel, will contribute its share of pollution during the combustion process. Particularly prominent is the discharge of sulfur dioxide, an eye, skin, and mucous membrane irritant, which is best known today for its connection with acid rain.

In summary, it is better for susceptible people to seek cleaner means of generating heat for daily living by employing electric and/or solar devices. There is no use denying it. For health-conscious individuals, natural gas—and other heating fuels—just aren't so hot.

4

THE SERIOUSNESS OF SMOKING

It all started in this country when Christopher Columbus discovered the Indian's use of tobacco. Not only did they possess the idea that it held medicinal value, but it played an important role in rituals such as the celebrated smoking of the peace pipe. Eventually, smoking was adopted, not only by the pioneers, but by the rest of the world as well. In recent times, statistics show that roughly one-third of the population smoke cigarettes, and a major study on health effects of air pollution has revealed that at least one smoker resides in some 70% of the homes in eight target cities.

Unfortunately, for the multibillion-dollar tobacco industry, there is just nothing good that can be said about cigarette smoking. No matter how insistent a person is to the contrary, no matter what one wants to believe, the facts are in—and none of them are favorable. Neither the medical profession nor any other scientific organization in existence today can state otherwise. It is quite probably the largest source of indoor air pollution. Every year, more people die from conditions attributable to smoking than perish from all other single health hazards combined—including automobile accidents. That's more than the number of Americans that lost their lives in every war this nation has fought in the twentieth century.

Constituents of Tobacco Smoke

Tobacco smoke contains, not a handful, not dozens, but in the neighborhood of an unbelievable 1500 to 3000 chemicals and other pollutants—some of which remain in the lung of the smoker and gradually accumulate, while a few are actually absorbed through the inner linings of the mouth, nose, and pharynx. Besides the well-known tar and nicotine, there are such harmful substances as carbon monoxide, benzene, pyrene, benzo-a-pyrene, fluoranthene, nitrogen dioxide, aldehydes (including formaldehyde), methane, cadmium, lead, phenols, ammonia, aluminum, sulfur, hydrogen cyanide, hydrogen sulfide, hydrocyanic acid, carbon disulfide, acrolein (the basic constituent of tear gas), polonium-210, toluene, pyridine, and even heavy dust. (See Table 2 in Appendix B.)

Additionally, one must consider that since tobacco is initially a crop, it can suffer from diseases and be plagued by pests. One of the means of controlling these problems is—you guessed it—by the use of sprays (including DDT, a percentage of which still endures in the soil, although the product was banned in the United States in 1972). Even the stored leaf is not invulnerable to insect attack and it is often protected by fumigants. Residues of these substances generally remain on the product to be inhaled along with the other harmful mixtures.

Besides that, the curing process usually requires the use of heating fuel, which can be in the from of wood, coal, coke, charcoal, oil, kerosene, or liquid petroleum gas. Regardless of which kind is used, some amount of the combustible remains are absorbed by the leaf. Once the product captures a dose of it, it can never be removed—even if there is an attempt made to do so.

There are also many additives in the way of flavorings and sweeteners. (Tobacco is otherwise tasteless.) While that doesn't sound particularly threatening, be it noted that they are composed of synthetic chemicals. Moreover, the tobacco companies refuse to reveal their exact nature.

Actually, the problem that any one of these numerous chemicals poses every time one chooses to light up is bad enough. A most noteworthy example is carbon monoxide, the chief constituent of auto exhaust. It alone makes up 1% to 5% of the smoke, generally more than any of the other components. It's inconceivable that anyone would spend various moments during the day behind their idling car. Yet that, in essence, is what one is doing by refusing to cast aside the cigarettes.

The inhalation of carbon monoxide impedes the body's ability to distribute oxygen because the blood cells absorb it approximately 10 times faster than the oxygen itself. The heart and lungs must work harder to get the oxygen it needs. Therefore respiration and heart rate increase. Blood pressure often rises. The body is under more strain and can suffer from headaches, fatigue, dizziness, nausea, loss of appetite, irregular heart beats, visual disturbances, confusion, personality changes, and impaired coordination, which can affect one's driving ability. In addition, it can hasten hardening of the arteries.

If that's not sufficient enough in itself to convince one to kick the habit, consider this fact. The concentration of hydrogen cyanide in cigarette smoke is 1600 ppm. Long-term exposure in the amount of approximately 10 ppm of this potent poison is considered dangerous.

And try this one on for size. Nicotine, the chief active principle of tobacco, is a highly poisonous liquid that is used as an *insecticide*! Just check the dictionary.

Acetaldehyde (an aldehyde often used as a solvent) is also present in the smoke. From the very first moment a puff is inhaled, this toxin combines with body protein and tends to create a stiffening condition in the connective tissues.

If that still isn't enough, tar, the particulate matter that remains after the moisture and nicotine have been subtracted, consists mainly of polycyclic aromatic hydrocarbons (PAH). Many of these are proven carcinogens. And just one cigarette can release in excess of 30 milligrams of tar. Benzo-a-pyrene is also a known carcinogen, as is benzene (the substance responsible for the odor of rubber cement) which has been linked to anemia, bone marrow damage, chromosome deterioration, and leukemia, not to mention the role it plays as a liver, kidney, and gastrointestinal tract irritant. Also, everyone by now knows the dangers of lead and DDT, not to mention combustibles such as nitrogen dioxide, which was brought out in the previous chapter.

In addition to all of this, one known tobacco additive is tung oil (or wood oil). This poisonous substance is used in paints and varnishes, and is thought to be connected to Chronic Epstein Barr Virus syndrome (more commonly known as Chronic Fatigue syndrome or "yuppie flu"), which is the causative agent of infectious mononucleosis.[1]

Furthermore, a recent finding based on research with mice at Argonne National Laboratory near Chicago, suggests that cadmium in tobacco, which accumulates in the liver and kidneys and is suspected of causing lung and prostate cancer, may promote osteoporosis.[2]

Sidestream Smoke

It is easy to see that this incredible barrage is murder on a smoker's bodily defenses. But it doesn't stop there. Sadly, even the nonsmokers—and ex-smokers—are not beyond the range of these noxious, deadly chemicals. In fact, the situation can actually be worse for anyone who does not wish to smoke, but must endure the burning pollutants of tobacco within an enclosed area. Since there is no filter on the other end of the cigarette, it only stands to reason that these emissions (known as sidestream smoke) are more potent. Dr. Zamm tells us that, "smoke from a cigarette burning in an ashtray contains almost twice the tar and nicotine of smoke inhaled from a cigarette, and so may be twice as toxic as smoke inhaled by the smoker."

Another consideration is the volume of smoke produced. The smoker may inhale 8 to 10 times per cigarette, but the sidestream smoke is being generated the entire time the cigarette burns.

It is an interesting fact, too, that the hotter the cigarette burns, the faster the combustible by-products will be released, and the more potent they will be. Most cigarettes burn at a rate of over 1600° Fahrenheit. The cheaper tobaccos generate greater heat.

Actually, the harmful concentrations that cigarette smoke can produce in a closed room far exceed the level of many industrial toxic substances that are now under government control. Benzo-a-pyrene has been measured at a rate of 10 to 30 times that of the outside air. Even with ventilation and air filtration, the inside concentration of carbon monoxide is well above the National Ambient Air Quality Standard of 9 ppm. Heavy smoking combined with poor ventilation has been known to generate a level in excess of 4 times that. Moreover, studies have revealed blood nicotine levels in nonsmokers exposed to such volumes of smoke to be as high as 20% of those that are found in smokers.

Of course, chemicals aren't the only problem that cigarette smoke produces abundantly in cozy confines. Even the level of particulate matter can be 10 to 100 times greater than the acceptable limits set for outdoor air. This is not only because the smoke contributes its own content to the air, but because it tends to gather up other particles and keeps them suspended for long periods of time.

And lest he or she not forget, the cigarette user, in addition to inhaling the filtered smoke (mainstream smoke), takes in the sidestream smoke too, as well as rebreathing some of the exhaled smoke (second-hand smoke).

Cigars and pipes are just as much a menace, although as long as a smoker doesn't choose to inhale the smoke, he does himself a favor. Actually, they are worse for the smoker than cigarettes if he *does* inhale because of the absence of a filter. However, he must also remember that, just as in the case of cigarettes, even if he doesn't inhale, he is going to be breathing a certain percentage of the sidestream and second-hand smoke anyway. And as far as sidestream smoke is concerned, a 1977 report from the American Lung Association reveals that a greater volume is produced by these products, which in turn generates higher levels of carbon monoxide, phenol, and benzo-a-pyrene.

And it is appropriate to add that tobacco doesn't have to be in its burning state to do harm. It is a toxic substance, period. Chewing it can cause cancer of the mouth, throat, and digestive tract. Using it as snuff can result in gum disease and tooth abrasion, as well as throat and mouth cancer.

Sobering Statistics

There are countless statistics that have been compiled regarding the harmful effects of cigarette smoking. At least some are worthy of mention.

■ Cigarette smoking is a major factor in the approximately 345,000 annual deaths from cancer and from heart, lung, and circulatory system diseases.

■ In the past 40 years, there has been an increased incidence of disease in women which coincides with their substantial increase in adopting the smoking habit.

■ A study conducted at Edinboro State College in Pennsylvania showed that nonsmoking women whose husbands did not smoke lived about 4 years longer than those whose husbands *did* smoke.

■ Another study revealed that for children under 5, the risk of upper respiratory tract infection doubles if their mother smokes.

■ In yet another study, 9 out of 10 asthmatic children improved dramatically when their parents quit smoking.

■ The risk of coronary artery disease is approximately 60% to 70% higher in men who smoke than those who don't.

■ Compared to nonsmokers, men who smoke just one pack of cigarettes a day increase their risk of contracting lung cancer 10-fold.

■ 85% to 90% of the deaths from lung cancer in 1980 was linked to smoking.

■ According to data garnered by the AMA's educational research foundation, women who smoke are 5 times as likely to develop lung cancer than those who don't smoke.

■ Cigarette smoking is the leading contributory cause of death from chronic bronchitis and emphysema.

■ The World Health Organization in 1976 revealed that the risk of esophagus cancer in smokers was greater than 6 times that of nonsmokers.

■ The risk of developing cancer of the pancreas is some 5 times greater for a two-pack-a-day smoker than for a nonsmoker.

■ Smokers not only possess the tendency to catch more colds than nonsmokers, but smoking will lengthen the course of the infection.

■ Smokers are inclined to have an increased need for vitamins. Just 3 cigarettes, for instance, are enough to negate all the important benefits that are normally derived from the vitamin C in an average glass of orange juice.

■ Pregnant smokers are more likely to give premature deliveries and suffer stillbirths.

■ Mothers who smoke during lactation have a high nicotine content in their milk.

■ Smoking has been implicated in Sudden Infant Death Syndrome (SIDS).

■ Smoking can cause early menopause in addition to reducing the number of active sperm in males.

Other Health Effects of Smoking

The habit of smoking can also lead to periods of apprehensiveness, depression, and feelings of uncertainty. It's a most unfortunate irony, too, that such negative emotions actually bring on the urge for a another cigarette. Other symptoms caused by cigarette smoke are excessive belching, flatulence (gas), heartburn, and sinus congestion.

Tobacco smoke is also responsible for dullening the sense of taste and smell, and increasing cholesterol levels. In addition, the Surgeon General has determined that it can promote circulation problems, especially in the fingers and toes. This condition not only tends to age the skin due to the delivery of less oxygen, but it sometimes becomes serious and leads to thromboangiitis obliterans (Buerger's disease), which involves premature occlusive changes in the peripheral arteries that can result in gangrene. (It is interesting to note that this disease is rare in nonsmokers.)

And contrary to some beliefs, cigarette smoking does nothing to aid digestion. A percentage of some of the substances, such as nicotine and tar, are absorbed by the saliva and swallowed to irritate the gastric mucous membranes, which prompts an excess secretion of hydrochloric acid. This, in turn, can produce peptic ulcers. Even a few cigarettes are capable of substantially increasing stomach acidity.

Furthermore, cigarette smoke can cause deafness, as the delicate structure of the inner ear is easily damaged. In the same way, it eventually paralyzes and destroys the cilia in the lungs and trachea. That, of course, eliminates one of the body's defenses, making it more vulnerable not only to cigarette smoke, but to all other pollution as well. Smokers becoming lung cancer victims who have also received excessive doses of radon, for instance, will have a tendency to develop the disease faster and succumb to it at an younger age, as is the fact that the cancer risk is increased 90-fold in asbestos workers who smoke.[3] Furthermore, this weakened body condition not only increases the chances of more frequent colds, but also promotes such infectious diseases as tuberculosis.

Employers look at the problem another way. Cigarette smokers have compiled an absentee work record which is over 45% greater than that of the nonsmoker, and they use the health care system at least 50% more often. During their working years, smoking employees average more than twice the mortality rate of nonsmokers, and those same workers consume approximately 6% of that working time engaged in the process of lighting up, smoking, and extinguishing their cigarettes. In 1981, a professor of business administration at Seattle University, William L. Weiss, put a monetary figure on it. He cautioned that each smoking worker was costing employers over $5,600 annually. And it is believed that as much as 10% of that figure could be attributable to the health effects of nonsmoking employees exposed to sidestream and second-hand smoke.

In all, the harm tobacco has unleashed on society is staggering. It is simply too much for the body to fight. And the saddest part of all, virtually no one is unaware. The Surgeon General has had his say on the cigarette packages, TV ads on the subject have been banned, many public places now prohibit it, and even the tobacco companies themselves have taken steps in an attempt to minimize the situation by developing more efficient filters and a product with reduced amounts of tar, nicotine, and carbon monoxide. Moreover, the National Cancer Institute reports that most smokers, perhaps as many as 90%, have made an earnest attempt to quit at least once, or admit they would if they could find as easy way to break the habit.

A Habit That Can Be Broken

Although there is no simple answer, it is encouraging that many (more than 30 million Americans since 1964) have conquered their compulsion, which proves to the remaining smokers that it is clearly possible to do so. Perhaps the smoker would do well to recall how conditions were before ever having lit up the first cigarette. (After all, no one was born with the habit.) Did he or she feel as though it was a necessity then to indulge in order to keep going? The answer to that one is an unwavering "no," but shortly after the habit is developed, it becomes an inevitable "yes."

The highly addictive nicotine, of course, is the nemesis. Strong and quick-acting, it possesses the ability to stimulate the nervous system, heart, and other organs. Yet in doing so, it not only increases the heart rate, but raises blood pressure, contracts the blood vessels, and is suspected of increasing the stickiness of the platelets (tiny disks in the blood involved in blood clotting) which obstruct the vessels. It can also affect appetite, sleep, and body temperature, in addition to causing nausea, vomiting, diarrhea, headaches, and dizziness. If the dose is large enough, it can result in convulsions, irregular heartbeats, coma, or even sudden heart failure. Nicotine will also depress the respiratory system of fetuses, impairing their lung development. The tobacco in cigarettes can contain as much as 5% of this chemical. A smoker may take in as much as 2 milligrams per cigarette. (Sixty milligrams—a thimbleful—if taken all at once, could kill an adult.) After it is absorbed into the blood, it reaches the brain within eight seconds. However, once it is completely flushed from the human system, the desire for it is removed.

And there are rewards in store for the ones who succeed in breaking the smoking habit. The body begins to heal itself within 12 hours. There is a rapid drop in the carbon monoxide level in the blood, and the heart and lungs are free to start repairing the damages. Energy will be restored. A person will have less tendency to become winded. The sense of taste and smell return. Headaches and hacking coughs will likely become a thing of the past. Furthermore, an encouraging statistic is that after about 15 years or so of abstinence, the death rate for ex-smokers is hardly more than that of nonsmokers.

In addition, there will be no more yellow-stained teeth, tongue, and fingers; nor smelly hair and clothes. What there *will* be is a sense of personal satisfaction to accompany a much healthier body.

Of course, one will most likely experience withdrawal symptoms immediately after cessation. Intense craving, tension, restlessness, irritability, constipation, depression, difficulty in concentrating, and disruption of sleep are not uncommon. But they should be ignored. They are temporary setbacks, a sure sign of recovery. Patience will *always* pay dividends. After the first 3 or 4 days of cessation, the cravings for a smoke will have peaked, and from that time on will begin to decline sharply. Other symptoms will also diminish and soon disappear.

The prospect of gaining weight also exists. But it should not be feared. It has been determined that a two-pack-a-day smoker of average weight would have to gain at least 75 pounds to offset the improvement in life expectancy that would be gained from kicking the habit. It would be far more desirable to regain a healthy appetite and risk putting on a few pounds in order to rid yourself of the multitude of harmful substances that have been continually coursing their way through your body.

There are many helpful books on the subject. They include: *Smoking: Your Choice Between Life and Death* by Alton Ochsner, *Stop Smoking, Lose Weight* by Neil Solomon, and *Scientific Case Against Smoking* by Ruth Winter.

Meanwhile, here are some tips from the experts.

Number 1, don't get tricky and attempt to taper off by switching to cigarettes which are lower in tar and nicotine. You will only be kidding yourself. Tests have shown that smokers of these products have a tendency not only to inhale more deeply and/or take larger or more frequent puffs, but are inclined to actually consume more cigarettes, not less, because the body keeps demanding the same dosage. In fact, don't use the gradual approach under any circumstances. *Stop abruptly*. Like pulling off a bandage, it is always less painful to get on with it and get it over with.

Number 2, *don't depend on tranquilizers* to help you quit. There are no facts to support the view that this is a remedy.

Number 3, take some *time* to *analyze* why you smoke and attack the problem from there. Do you get a lift? Does it calm your nerves? Is it related more to social satisfaction? Or is it a combination of reasons? The more details you know about the problem, the easier it is going to be to eliminate it.

Number 4, *discard* items associated with smoking, such as ash trays, lighters, and cigarette holders.

Number 5, *adopt substitute habits*—obviously good ones and ones that bring you pleasure—like eating (as long as you don't get too carried away), beginning a new hobby, or making purchases of those fun items you've always wanted but never thought you should indulge in. (You'll be more able to afford any of these suggestions on the funds saved from eliminating cigarette purchases.)

Number 6, *encourage* others to stop, and *tell* them you've stopped. This can reinforce your efforts at the same time it is helping them.

Number 7, *think about your loved ones*, especially children. They deserve protection from harmful pollution. In addition, you will be setting a good example. What parent was ever eager to see an offspring start smoking?

Many individuals, often employing some or all of the above advice, have resolved to quit smoking on their own and succeeded. But if these last few pages have induced you into that long-delayed decision, the one you've really wanted to make all along anyway, and you should need support to get things rolling, any local public or private health agency can furnish information concerning community programs that are designed to assist you in licking this problem. Don't hesitate to inquire. It will truly be one of the wisest decisions you could ever make.

Certainly not everyone will, in times to come, give up smoking. But, basically, those who become fully convinced of the dangers, who finally realize they not only *can* do it, but *must* do it because their human system is not unlike everyone else's, will have found the motivation to do so. It still won't go without effort (no worthwhile pursuit ever does), but keep in mind that habits, just like promises, can be—and often are—broken. Will you be among those fortunate ones?

THE DUST DILEMMA

The most commonly thought of substance when considering dirty air is dust. And that is a valid association, for it is as abundant as water vapor. We live with this grayish, powdery material constantly and are always breathing it to some extent. There is no other choice unless we shift our residence to outer space.

If air is reasonably pure (such as atop a mountain retreat), it may contain some 500 particles of dust per cubic centimeter. (A cubic centimeter, or c.c., is equal to 0.06 of a cubic inch.) In dirty air (within a bustling city), that count can zoom to 50,000 particles per c.c. But when there is a considerable amount of activity inside a house or building within that bustling city, the concentration is often elevated to at least twice that.

Substances that Mix With Dust

House dust has been given a large share of the blame for incidents of rhinitis (inflammation of the nasal membranes). It has also been branded as an irritant to asthmatics. Many doctors have dubbed it an "occupational hazard" for the homemaker.

This substance really isn't as bad, though, as the particles riding on it. They consist of an incredible and diverse collection of matter which includes bacteria and viruses; mold spores; dust mites; bits of cotton, wool, wood, feathers, and hair (both human and animal); dander (both human and animal); sebum (both human and animal); pesticide powder; detergents; food remnants; insect fragments; insect excretions; shreds of kapok and cellulose; algae; residue from cosmetics; paint chips; bits of plaster; pollen; wallpaper flakes; and fireplace soot.

Dust mites (dermatophagoides) are probably the most associated with dust; and truly, they are a common component. One study group of entomologists collected dust from 64 homes in the eastern part of the country and found dust mites inhabiting nearly two-thirds of the samples.

These microscopic arthropods, which live in temperatures above 50° Fahrenheit, feed on the flakes of skin we regularly shed, as well as food particles, and are especially fond of humid conditions. Some of their favorite breeding grounds are upholstery and bedding in damp rooms. (Sweating and bedwetting can contribute to this.) They have been known to cause severe itch, and there are those who have developed an allergy to them and/or their excrements. It is possible, too, for them to aggravate other allergies. According to David Rousseau, author of *Your Home, Your Health, and Well-Being*, "their presence is the major reason why house dust causes reactions among allergy sufferers, even when pollens and house pets are not present."

And it may sound contradictory, but although dust mites flourish during the summer months, those affected may actually find their symptoms worse in winter. This is because by the winter season these tiny creatures have died, and their dried up bodies have broken down into fragments. These fragments, being obviously smaller than the full-sized mite, can more readily reach the respiratory tract.

Dander, most notably that which is shed by domestic animals, is, of course, highly allergenic, as is pollen. Both mix readily with dust in the home and have been known to keep some people in a constant state of misery.

The inhalation of insect fragments, too, can be troublesome for others. Perhaps this is still due to the dust, as well as pollen or other particles that the small bits of scales, wings, and bodies can contain, but no one yet knows for sure. (For suggestions on environmentally safe ways to keep insects under control, see Chapter 6, *The Chemical Crisis*.)

Pesticides, another important consideration, have been known to be transported long distances in the outside air on dust particles. Just think how much more easily they can be moved about in the confines of a house or other building.

There are a few bacteria that can live for remarkably long periods dried in dust, hence the reason dust has been pegged as one of the most common elements capable of transmitting disease. *Clostridium tetani* (tetanus) is one of the most frequent varieties present. (See Chapter 14, *What About Bacteria?*)

How Dust is Distributed

Large quantities of dust are commonly disturbed whenever portable fans are used, or when air conditioning or heating is first turned on at the beginning of a new season due to the build up around the vents and places in direct contact with the blowing air. Added to this are the particles that escape the filter. Heating systems can be worse because since warmer air is less dense, there are a smaller number of its molecules per cubic centimeter, and it therefore has a greater capacity to hold particles.

What's more, dust that gets hot enough can have a particularly bad effect on some individuals. The high operating temperature of the elements in an electric heater is capable of superheating airborne dust. So is that of an incandescent light bulb. This "cooking" process alters it chemically, making it even more potentially harmful.[1]

Sadly, it's the larger particles that have a tendency to settle first; the smaller, more easily breathable, remain suspended longer—and are more readily disturbed—because of their lighter weight.

Theoretically, every bit of matter would finally settle and stay out of circulation in a perfectly stationary setting, but what normal interior environment would not contain the hustle and bustle of daily activity?

Eliminating Dust

After considering all of this, it might almost be tempting to give outer space a whirl—especially since even the best of housekeepers is incapable of eradicating every bit of it. In fact, one study disclosed that an average six-room house in the city can accumulate as much as 40 pounds of dust annually.[2] One can get a good idea of just how dense this concentration is whenever eyeing a shaft of sunlight from the proper angle. It will highlight a section of the countless sea of specks defiantly riding the air currents. The typical housewife can attest to how bad the problem is every time she notices that the furniture surfaces are suspiciously dusty far too soon after house cleaning. And if that housewife always feels bad during or after her chore, she could very well be sensitive to dust, if not one or more of the substances using the dust as a vehicle to make its way into her respiratory passages.

Vacuum-cleaning Equipment

Although the situation can prove serious, there are ways to alleviate it—and it starts with the process of vacuuming. There is a gross misinterpretation about this. It may sound as surprising as a five-legged dog or as illogical as falling up a mountain, but the truth is, a conventional vacuum cleaner is no better than a mere toy for picking up dust.[3] Its only capability is to eat the largest particles of dirt and debris. The smaller bits of matter (especially in the 0.5 to 2 micron range)—that which can more easily enter the body—are utterly sucked up and blown back out through the porous bag (the only means of exhaust the device possesses) to be redistributed into the air.

So rather than being eliminated, the majority of the most troublesome particles are simply rearranged, placing them in the ideal position to do harm to health. Roger Fox, M. D., who is chairman of the American Academy of Allergy and Immunology's Committee on Environmental Controls and Indoor Air Pollution, tells us that the allergen content of the air can rise from 2 to 10 times that of what it was before vacuuming.

Anyone with doubts need only to turn on a conventional vacuum cleaner with a partially-full bag and look closely while

Air flow of conventional vacuum cleaner. A conventional vacuum cleaner draws in air from underneath and exhausts it (and many of the microscopic apticles it contains) through the porous bag.

inward flow of air

outward flow of air

shining a strong light on the exhaust grill (the place around the bag where the warm air is rushing out). It is amazing what one can often see being expelled before even running the device across a carpet.

A better test still is to completely cover one side of a small index card or heavy piece of paper with double stick tape, and attach it to a portion of the exhaust area, making sure it is secure by placing more tape along the outside borders. (Don't cover the entire portion of the exhaust or the air flow will be cut off.) Then try vacuuming part of a room. You won't believe what turns up on the tape.

As it is, the air flow on a conventional vacuum cleaner is weak anyway. Many dust particles are left behind only to be stirred up later by foot traffic. The more packed the bag gets, the less effective your cleaning job will be because the more the air flow will be reduced and, therefore, the less the suction will be. To get an idea of how much dirt remains on a carpet, fold a loose corner of it over a piece of fresh white paper and tap the backing. Chances are, your paper will not still be white.

The only defense available if an allergic person insists on using this approach, is to wear a dust mask, preferably the cotton surgical kind, which is the best choice of the economically-priced variety. Or if someone else (hopefully less susceptible) is doing the chore, it would be advisable for the more sensitive to vacate the premises for at least an hour to give most of the particles a chance to settle. It is also highly advisable to change the bag out-of-doors—always—even if one must wait for appropriate weather conditions. Additionally, it is better to replace the bag when only half full, to clean the area around the bag with a moist cloth before installing a fresh bag, and to run the machine for at least 30 seconds before returning it to the house so any loose dust that remains will be expelled.

But in reality, as far as conventional methods go, it would be just as well for a susceptible person to employ a non-electric portable sweeper (known to most as a carpet sweeper) and empty it carefully outside, or not bother to vacuum at all!

Fortunately, though, there are alternatives. Accessory filters are available that can be added to some conventional vacuum cleaners (the kind employing a hard outer shell), thereby reducing the number of particles they emit. They aren't designed to function on machines that use a soft bag, however. A pity, since the soft bag varieties are the worst offenders.

At any rate, the filters require regular change or their purpose will be defeated. If a blockage occurs out of neglect, debris can be forced through the sides of the filter or the mechanism could lose suction. Also, in some instances, the motor could overheat.

Another option involves the use of other types of vacuum cleaning equipment which perform considerably better. One is the central vacuum system, which employs a network of ducts throughout the house that directs dirt to a unit usually mounted in the garage. There is no question that it is a worthwhile device. At least it deposits the dirt somewhere else besides the interior of the house. But installing all the duct work that would be required can be impractical as well as expensive.

A number of other portable varieties are on the market which use a bag and filter combination that are capable of removing particles in the 0.3 to 0.5 micron range. That far outpaces the conventional kind, but one still has to face the prospect of replacing the bag and changing the filters sooner or later. The tiny dry matter which is readily liberated into the air during this process can prove disastrous to sensitive individuals. Again, replacement of the bag and filter should be done outside, and it is advisable to wear a dust mask when doing so.

The best bet seems to be the kind of portable unit whose operation depends on a basin of water to trap the dirt. In this type of system, known to some as a water-trap vacuum cleaner, particle size is academic. Any and all water-soluble material that is pulled in is absorbed by the swirling liquid. Air flow is never impeded because since the water makes the device so efficient, there is no filter to become clogged. Furthermore, only pure air is released from the exhaust ports. Also, the emptying process is hardly threatening health-wise because one is dealing with dirty water—not the dry matter that has such a great affinity for the air. Furthermore, since this is a water-base system, it can be used for wet pick up as well—in all a must for any individual who is affected by dust. (See Appendix G for information regarding this type of

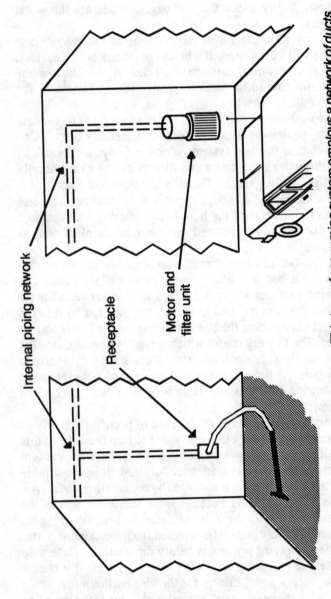

Depiction of a central vacuum system. This type of vacuuming system employs a network of ducts throughout the house that directs dirt to a motor and filter unit located outside the living quarters.

Internal piping network

Receptacle

Motor and filter unit

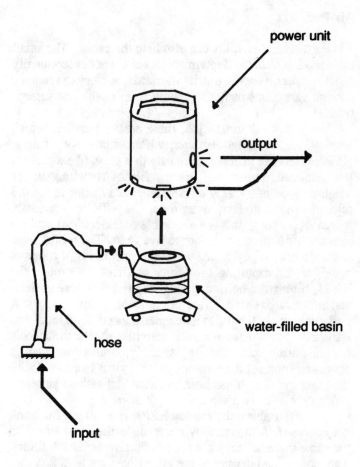

Components of a water-trap vacuuming system. Hose attaches to water basin; water basin locks onto power unit. Air is drawn into hose, through basin of swirling water (where particulate matter is captured), and out holes in power unit.

vacuuming system.) For information concerning vacuum cleaning devices and equipment in general, check "Vacuum Cleaners" in the yellow pages.

Air Purifiers

Portable air purifiers can also help the cause.[4] The small table models found in department stores are not exceptionally efficient when it comes to strict standards, but they can remove enough particulate matter to benefit some of the less susceptible.

For a more thorough job, there are air purifiers which employ HEPA (High Efficiency Particulate Air) filters. HEPA filters are made of extremely thin glass fibers pressed into a pleated paper, and are capable of removing in the neighborhood of 99.97% of the particulate matter in the 0.3 micron range. In fact, over time, the efficiency actually increases (although this is a negligible consideration), and the average life expectancy is four to five years. The HEPA filter is customarily piggybacked with two other kinds of filters: a pre-filter for capturing the larger particles (to protect the HEPA filter from becoming clogged, and therefore useless) and an activated charcoal filter for absorbing certain odors. A portable unit containing these elements meets rigid industrial standards and is used for such activities as electronic chip manufacturing and satellite assembly, and is utilized in operating rooms, microbiological labs, virus isolation facilities, and the like. It has been reported that asthma sufferers profit dramatically when using such devices.

Some portable units employ HEPA-*type* filters. But don't get confused. Although they are made in the same way with the same material, they are not as effective as HEPA filters. The number and density of the glass fibers are less, and they are generally only about 95% efficient at 3 microns (not 0.3). In addition to HEPA-type portable models, there are HEPA-type filters available for installation in the return air duct of central heating and air conditioning systems—still a much better choice than a conventional filter.

Also, in lieu of, or in addition to any of the above measures, one can consider the installation of charcoal filter pads to each of the room ducts of central air systems.

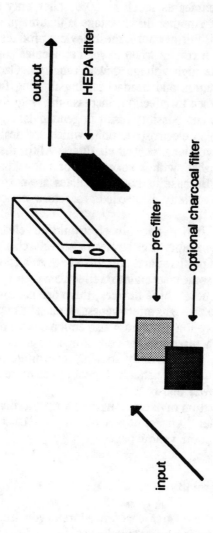

Components of a typical HEPA filtration system. Air is drawn in by electric motor through charcoal and pre-filter, and forced out through HEPA filter.

input

optional charcoal filter

pre-filter

output

HEPA filter

Electronic (or electrostatic) air cleaners are another option. These portable units charge particles electrically and pull them out of the air, where they are deposited on a special plate. They are not nearly as efficient, however, as HEPA filtration, and tests have proven that that efficiency actually drops substantially (sometimes as much as 60%) after only a short period of usage. Another disadvantage is that the plate must be cleaned often. Furthermore, these devices produce ozone, a pollutant which results when oxygen molecules come into contact with the high voltage that is required. Ozone can induce eye irritation and headaches, impair lung function, diminish resistance to infection, increase the fragility of the red blood cells, and possibly result in genetic damage.

There are also electrostatic units which are designed to replace the conventional central air filters. All of these have the same drawbacks, with the exception of a new kind called an "electret." Because its collector plates are permanently charged, no electricity is needed, therefore no ozone is produced.

A similar kind of portable electronic air cleaner is a negative-ion generator. It generates negatively charged ions which intercept particles and pull them back to a filter in the unit. Although some have claimed relief from their symptoms by the use of one of these devices, there are no convincing statistics to help prove its effectiveness. It must be mentioned, too, that negative-ion generators also emit a degree of ozone.

The manufacturers of these electronic air cleaning products, however, are constantly striving to institute improvements. In the future, we should expect to see more efficient devices with fewer disadvantages.

For information on the purchase of air purifying apparatuses, look under "Air Pollution Control" or "Dust Collecting Systems" in the yellow pages.

Control of Humidity

In any case, no matter what procedure(s) you decide to use to control dust, it is advisable to keep the humidity level low. Not only does it offer maximum comfort to the body, but it discourages the proliferation of bacteria, mold, and dust mites. Besides, dust has a habit of clinging to moisture in the air. If your home is not equipped with a dehumidifier (either as part of a central air system or as a portable unit), by all means obtain one and use it if humidity is a problem. Make sure, though, that it is supplied with a tube or piping to divert moisture out-of-doors or it will be necessary to drain and clean it daily to prevent it, itself, from accumulating bacteria and mold.

This warning, however: If the humidity level gets too low, it can not only play its part in encouraging dust to become airborne, but it will create a drying effect upon the protective linings of the nose, throat, and lungs. Some medical researchers are of the opinion this condition retards the body's ability to combat microorganisms. But the verdict on this is not yet in.

At any rate, a relative humidity in the range of 40% to 60% is recommended by the experts. Hygrometers, devices that measure this condition, are inexpensive and readily available.

Another tip is to maintain a lower than average temperature. While this is no substitute for cleaning, it will at least help somewhat in keeping down the airborne particle count.

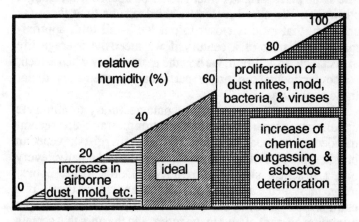

Relative humidity's effect on the indoor environment. *Although a dry atmosphere is capable of allowing more particulate matter into the air, excessive humidity will encourage the survival of several potentially-harmful organisms as well as increase asbestos deterioration and the outgassing of chemicals such as formaldehyde.* Source: Adapted from Rousseau, David, *Your Home, Your Health, and Well-Being,* (Vancouver, B. C., Hartley & Marks, 1988.)

Saying Good-bye to "Dust-catchers"

There are also a number of other less expensive measures for the dust-sensitive person. First, rid yourself of the broom habit. Sweeping can stir up contaminants just like that of a conventional vacuum cleaner. In addition, a lot of these particles remain on the broom itself, some to be released by air currents whenever the closet door is opened, for instance, or perhaps liberated (along with many of the other particles that are being swept) the next time one puts the broom into action. A better bet would be to use cloths and mops moistened with water (not chemicals, which create their own brand of pollution). At least that will keep most of the dust confined to the cleaning articles until they can be rinsed out.

Secondly, one can remove many of the things that attract accumulations of dust—little knickknacks and odd-shaped items difficult to dust around. It never costs anything to discard, and there are a number of things one can part with—especially if health is involved.

The first place to begin such a campaign is the bedroom (the same place a HEPA filter should be housed if the budget allows it). We spend at least a third of our time there (as children, that ratio is closer to one half)—all told, approximately a quarter of a century if we meet the average life expectancy. In addition, we breathe more deeply when asleep, so contaminants can prove particularly penetrating under these conditions.

In extreme cases, it may become necessary to eliminate upholstery, draperies (washable cotton curtains are recommended as a good substitute by most allergists), venetian blinds, and carpeting (the worst offender). Actually, every object practical should be removed. Don't forget books, magazines, bed canopies, and even hanging pictures. If a child's room is involved, it would be a good idea to keep a minimum amount of toys in the room, and the ones that remain should be of the washable variety. No stuffed toys. Box springs and mattresses should be covered in a dust-proof casing—and nothing should ever be stored under the bed. Bunk beds should also be avoided.

It is important, too, that the bedroom closets only be used for clothing currently in use, and that it be hung and sealed in garment bags. All outer garments like coats and sweaters are best relegated to less frequently visited rooms, as are other objects normally appropriate for a bedroom closet. Of course, make sure the attire, as well as bedspreads, covers, sheets, and curtains, are washed regularly. It is surprising the amount of particulate matter they pick up. All one has to do is examine the dryer filter contents from a single load of laundry to be convinced. Using tightly-woven fabrics will help the cause. At least they will not invite as much dust. Also, make it a habit to clean walls, baseboards, door and door frame tops, and windows, preferably with plain soap and water.

Don't neglect the rest of the house, either. There are still many dust-catchers that can be eliminated, or if nothing better, demoted to a position in the garage (although an attached garage will invite some particulate matter back into the house). Keep in mind, too, that it would be better to forsake all carpeting and rugs than to allow any family member to go on suffering. They must absorb everything that is dished out to them on shoes, pets, and other objects. That's plenty of material to be stirred up every time someone walks by. (See Chapter 8, *Construction Materials that can Demolish Health*, for more details about carpeting.) If nothing better, at least contract to have carpeting, as well as upholstery steam cleaned occasionally. Remember, just because these objects don't show the dust like your coffee table doesn't mean it isn't there.

Other Items that Accumulate Dust

Other areas often passed over when dusting are refrigerator tops, ceiling light fixtures, the ornate carvings of the more intricate varieties of furniture, upper shelves, and closets. Even the top portion of ceiling fans require attention if they have been idle for a while. To make matters easier, hanging light fixtures, especially chandeliers, can be replaced with recessed lighting. That will eliminate some less accessible areas where dust unceasingly accumulates. In regard to closets, an amazing amount of dust can find its way inside even when the doors have been tightly shut. Don't crowd them with too many possessions. Also, consider the use of a wire shoe rack. When cleaning closet floors, it can be much more quickly removed than individual pairs of shoes.

Exhaust fans, too, are often overlooked. They must be cleaned occasionally or they cannot be relied upon to adequately remove dusty air. The cover plate can be removed and placed in hot soapy water. While it's soaking in the sink, wipe out the exhaust opening with a solution of borax and water. The motor unit can also be removed and brushed off, but don't expose it to water, as this could prove harmful to the mechanism.

Also, keep up with conventional air conditioning and heating filters by frequent replacement or cleaning, and have the ducts professionally cleaned each season if possible. They can easily remain one of the most contaminated portions of the house because they are so inaccessible for cleaning and such a large percentage of the matter in the air circulates through them. If you have electric baseboard heaters, remove the protective cover and clean the inside of the mechanism thoroughly. These units have a tendency to collect a lot of dust, as do wall radiators, which also need attention. Keep in mind, too, that these devices can be covered in the summertime, thereby protecting them from dust and other particulate matter.

One worthwhile addition to the home is that of the door mat—a good weapon with which to fight dust. Mats should be large, and one should be placed at every entry point, including the garage, where concrete dust can be a problem. Encouraging each member of the family to use them is a must. Of course, don't forget to clean them regularly.

And everyone may think you have become a fussy nuisance, but an even better idea is to request that shoes be removed before entering the premises. This is an excellent measure to see that the minimum amount of dust is carried inside.

In summary, just remember that the more action you take to eliminate dust, the less of it, as well as other particulate nasties, there will be floating in the air to threaten your health.

THE CHEMICAL CRISIS

A number of chemicals have already been discussed that frequently penetrate the air of our homes and of other buildings. But there are numerous chemical products we use everyday, largely taken for granted, which, besides being potential fire hazards and/or providing the possibility of poisoning by ingestion, add to indoor contamination considerably. Cleaners, pesticides, and cosmetics are just a few examples. And to complicate matters, the law does not require a complete list of ingredients on the contents label of some of these products because they are regarded as trade secrets.

Cleaning Chemicals

Cleaning chemicals are often one of the worst offenders—a harsh irony, since they are used in an effort to clean up dirt and other contaminants, not contribute to them. Because of the job they have been designed to do, they seem the most innocuous of any substances. Yet, as with most products, we tend to use them not only regularly, but liberally—and they contain some strong ingredients, such as ammonia, bleach, chlorine, phosphates, and sometimes volatile petroleum solvents in the form of acetone, benzene, naptha, toluene, and xylene. The dangers of such chemicals are becoming better known, as many have suffered a wide range of physical and mental consequences due to prolonged or frequent exposure.

In some instances, children who don't get along well at school turn out to be nothing more than the victims of janitorial supplies, which have been extracted from a broom closet and indiscriminately dispensed in the halls and the classrooms. Many cases of inattention, irritability, hyperactivity, and misbehavior have been linked to commercial cleaners, disinfectants, and the like, that are common around schools.[1]

One of the ingredients commonly found in disinfectants is cresol. This chemical attacks not only the liver, kidneys, spleen, and pancreas, but the central nervous system as well. A startling example of what can result from such an exposure involved a patient of Dr. Mandell's, a 10-year-old girl whose chemical sensitivity became so severe, that on one occasion, a disinfectant used in a school lavatory caused her to behave such that the teachers thought she had been taking drugs.

Of course, it's not uncommon for stores, malls, offices, and other public buildings to harbor the fumes from these products. And this is a sobering thought, but even hospitals, regarded as places of recuperation, are not off limits to such potent chemicals.

Aerosols

Aerosols can be especially bad, not just because of the solvents and volatile chemical mixtures, but also because of the propellants that must be used to dispatch these chemicals. Typical propellants are propane, butane, isobutane, pentane, nitrous oxide, and methylene chloride. (Freon must also be included in this list. It is a combination of carbon, fluorine, and often chlorine which, in large doses, has been known to cause coma and death. It has been discontinued because of its threat to the ozone layer but, undoubtedly, there are still a number of unused or partially-used products containing this substance, sitting innocently on a shelf or in a cabinet somewhere.) A significant portion of such chemicals, as well as the other contents, are released to evaporate into the air every time the product is used. It is estimated that as much as 15 pounds of propellant is released in the average household annually.[2]

At the very least, propellants can be an eye, nose, or throat irritant. Moreover, besides their obvious potential for creating respiratory problems, some have been known to cause hives and damage the cornea of the eye. In addition, research indicates that most of them can affect the heart as well as the

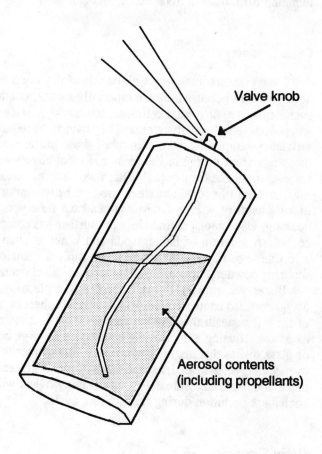

Valve knob

Aerosol contents
(including propellants)

A peek inside an aerosol container. Aerosol products despense potentially harmful chemical vapors into the air, including not only those required for the application intended, but those used for the propellants as well.

central nervous system (possibly resulting in brain cell damage), and that methylene chloride not only can convert to carbon monoxide when inhaled, but that it is capable of inducing birth defects, fetal death, and cancer.

Oven Cleaners

Oven cleaners may very well be one of the worst household polluters because they are especially strong, containing such items as hydroxethyl cellulose, sodium hydroxide (lye), and polyoxyethylene fatty ethers. They have to be to have the efficiency required to take care of all those burned-on food drippings from meat sauces, pies, and what-have-you, that have accumulated for months. What's worse, they release the most amount of propellant when used. A better option is to make a habit of wiping out your oven each time you finish cooking. Use a moist cloth before the surface has completely cooled. A solution of baking soda and water is even more effective. For more stubborn residue, apply a chlorine-free scouring powder, allow to dry, then scrub with a cleaning pad.

Better yet, prevent the spills in the first place by using adequate-sized containers, as well as a cookie sheet or a piece of tin foil beneath the cooking food in the event their is an overflow. If using a microwave oven, stick to large ceramic or glass dishes that are microwave safe. Remember that any oven spills, when heated, are sad news for the inside environment, too. And unfortunately, self-cleaning ovens will also contribute pollution during the cleaning cycle.

Drain Cleaners

Drain cleaners are also an important environmental concern. They are composed almost entirely of sodium hydroxide, which is highly toxic and is capable of irritating mucus membranes. Other ingredients include trichloroethane, petroleum distillates, and sometimes ammonia. (Warning: Never use a chlorine-based bleach or cleanser in conjunction with a drain cleaner containing ammonia. This combination will release chloramine, a deadly gas.)

Actually, the best policy when facing stopped up plumbing is to use a plunger. If that fails, removing and cleaning the drain trap may be necessary. Even the services of a plumber, if required, beats the use of strong chemicals. To prevent clogs, keep grease and larger particles out, and as an extra precaution, pour boiling water down the drain at least twice a week.

Metal Cleaners

Metal cleaners present a toxic problem, too. They should only be used outdoors and care should be taken that the fumes are never inhaled. A safer cleaner is a simple salt and vinegar mixture. Also, baking soda or cream of tartar in warm water is effective. Either formula should take care of brass, silver, or copper.

Clothes Detergents

Some have even been bothered by clothes detergents. Phosphates or other chemicals can leave residue in garments and will be breathed by the wearer all day long. Unless there are stubborn stains to contend with, borax is nearly always a safe substitute for clothes detergent. It not only cleans, but also deodorizes, disinfects, and whitens, as well as softening the water.

Generally, as far as cleaning chemicals go, the best policy is to substitute plain soap and water for the less demanding applications. Bar or flake soap without additives is recommended (although borax may have to be added in hard-water areas to enhance the effectness of the soap). If more abrasion is required, apply plain table salt.

Whatever cleaners you must use, keep them tightly sealed to minimize escaping vapors, and take the additional precaution of storing them in a shed or garage. (Please note, though, that a garage may not offer full protection if it is directly accessible to the house. Minute amounts of vapors can still escape into the living quarters through a doorway, even if it remains closed, or through imperfections in a wall to menace unusually sensitive individuals. Only detached garages are an acceptable storeroom in that case.)

Air Fresheners

Air fresheners rank as another major offender. Whether aerosol or ornamental, their job is to rid the air of bad odors. And that they do—but in the process, not only do they cover up warning signals of potential health hazards, they add their own chemical contaminants to the air, among which can include cresol, ethanol (ethyl alcohol), propylene glycol morpholine, and formaldehyde.[3] The key here is to simply seek out the source of the odors. If you eliminate the offending substance(s), the odor(s) will go with it. In the event a deodorizer must be used for lingering problems, nothing beats ordinary baking soda. It absorbs odors rather than adding its own—and it's cheaper.

Furniture Polish

Furniture polish is another widely-used item that deserves a word. It can contain things like chlorine, phenols, acrylics, and even formaldehyde and pesticides. These substances can emit vapors for lengthy periods of time. Melted beeswax and mineral oil are far more acceptable for polishing any kind of wood. Olive or vegetable oil and lemon juice are also effective.

Floor Wax

Floor wax also contains many of the same ingredients and other volatile chemicals. The safest product for this job is a paste wax. Although it contains petroleum solvents, they evaporate quickly, therefore making this kind of wax more chemically stable after it has dried than liquid varieties. Still, it may also contain resins which might prove unacceptable for some. Always use it, like any other chemical product, only when the area is properly ventilated.

Pesticides

Pesticides certainly deserve deep concern when it comes to the indoor environment. Although the dangers have been more widely publicized the last few years, and containers clearly display cautions regarding their application, many are still far too casual about the way they dispense these products. Even when extreme care is taken, it is often not enough. They may be used to a lesser extent than cleaning chemicals, but we must remember they are outright poisons. What's worse, once they have been applied, it is very difficult, and sometimes impossible, to completely remove them. There have been those who have given their homes a complete scrub down specifically for purposes of removing pesticides, only to learn later that residues were still present. In fact, extensive testing has turned up contamination in homes that were treated with the pesticides chlordane and heptachlor as much as 20 years ago![4]

Just a few of the countless active ingredients that are found in household pesticides are aldrin, dieldrin, toxaphene, mercury, dinitrophenol, bendiocarb, phenol methylcarbonate, dichlorvos, cyclopropane, toluamide, lindane, and arsenic compounds. These words are just as bad as they sound. Pesticides are designed to *kill*, whether it be insects (insecticides), rodents (rodenticide), mites (miticides), weeds, (herbicides), or molds (fungicides). And since we, like they, are composed of living cells, we stand to be affected too. That's not even considering the consequences of the inert ingredients, either. Although those substances are not disclosed on the label, they can actually be more toxic to the body than the active ingredients.

The result: a possibility of many physical and mental symptoms, including the not surprising respiratory difficulties. Dr. Bambi Young, Director of Environment and Behavior, Center for Science in the Public Interest, cites such potential problems as blurred vision, twitching, loss of coordination, mental confusion, digestive disturbances, weakness, dizziness, headaches, and tiredness. Other findings have also linked pesticide exposure to depression, memory loss, anxiety, paranoia, and social withdrawal. Some pesticides are suspected carcinogens, while still others are thought to cause Parkinson's Disease.

There have been many reported cases involving both children and adults alike who have suffered serious or fatal consequences just from standing at length on lawns freshly treated with pesticides.[5] One of the most horrendous examples of just how potent pesticides can be and what they can cause in this kind of circumstance, involved a healthy Air Force pilot who, in 1982, decided to play a game of golf on a course in Arlington, Virginia, which had been sprayed with a fungicide. He soon began suffering from flu-like symptoms which were accompanied by a headache, a rash, and a high fever. By the next day, his body was covered with severe blisters, and after two weeks he was comatose. Before another week, most of his skin had come off, and he died of kidney failure, and pneumonia.[6] If such incidents can take place in the open air, just think what can happen inside a home or public building where pesticides have been used.

After considering the above, you might deem yourself safe if you make it a policy to abandon all store-bought spray and powdered pesticides and refrain from ever calling an exterminator again, but chances are there is something you've overlooked. Step over to the nearest clothes closet, open the door, and take a whiff. Likely you will catch the scent of moth balls. This is more sad news for the home environment. They are made up of paradichlorobenzene and naphthalene. According to environmental engineer Francis Silver, "These chemicals I would estimate to produce more injury than is produced by automobile accidents." Among other things, they have been the cause of headaches, rapid heartbeats, depression, cataracts, high blood pressure, and liver and kidney ailments.

Obscure Examples of Insecticides

A red flag should even be waved in the direction of pest strips, as well as flea and tick collars and powders. They all contain a wide variety of insecticides. In the case of strips and collars, these insecticides are slowly released from the plastic to permeate the air. The administration of powder, of course, can dispense its share of toxic particulate matter. This kind of treatment should always be done out-of-doors, and by rights, the pet should remain there too.

Lawn and Garden Pesticides

Of course, take care when handling lawn and garden pesticides. Wear a charcoal filter mask, and make sure to dispense the product on a windless day, or at least one during which the wind isn't blowing strongly. If there is a breeze, make sure it is moving in a favorable direction. Otherwise, you might wind up with a dose of it in your house or a neighbor's, as they can always find entry through poorly-fitted doors and windows, attic vents, or perhaps other air inlets. Particulate matter and vapors can also blow back at you and land on clothes, shoes, skin, and hair to later be brought inside.

Also remember to properly seal all pesticides and store them outside the confines of the house, in their rightful container, preferably in a shed or a detached garage. If residing in an apartment, one might consider a weatherproof box, which can be placed on a balcony or back porch.

Insecticide Substitutes

It would be far more beneficial health-wise, however, to banish the majority of these virulent chemical mixtures to the trash can. There are safe, effective substitutes for handling many pests. The following are suggested:

■ For ants, red pepper can be used. Just sprinkle in strategic spots. Also, honey and boric acid or borax applied to small strips of paper will work. Another idea is to plant peppermint or pansies around doors and windows outside. They serve as good repellents.

■ To prevent beetles from getting into foodstuffs, store all vulnerable items, such as grains and flours, in a cool spot in tightly-sealed glass containers. Place a bay leaf in each container.

■ Brewer's yeast can be added to your pet's food to discourage fleas. Additionally, hot water and a mild, non-phosphate, biodegradable soap should be used to wash animal bedding frequently. Pine needles placed in and around dog houses also work for fleas. Another effective measure is to salt the crevices of the dog house. And remember not to neglect your pet's nutritional requirements. If a dog or cat should fall ill, it will be more prone to attack by fleas.

■ As far as moths are concerned, wash or brush infrequently-worn clothes regularly. This insect and its eggs are fragile and cannot take activity. For apparel that is to be stored for lengthy periods, wrap in paper and freeze for about a week prior to storing, then store in tightly-fitting bags. If a moth ball substitute is desired, try cedar wood shavings, blocks, or balls.

■ Boric acid is effective for roaches and silverfish. For roaches, sprinkle in corners, along baseboards, or wherever they are noticed. When dealing with silverfish, since they have such a remarkable affinity for paper, mix the boric acid with flour and sugar and place on strips of paper. Also, roach traps that contain natural food bait and no insecticide can be used not only for roaches, but for a number of other crawling insects. In addition, plain baking soda is toxic to roaches.

■ Basil plants grown around doors and windows will help repel flies. If necessary, hang flypaper (which can also be made by applying honey to paper) or use outdoor "electrocuters."

■ The breeding of mosquitos should be prevented by emptying any stagnant water on the premises. But if they are still a problem, grow basil plant around doors and windows or use outdoor "electrocuters."

■ If lice become a bother for dogs or cats, the adult insects can be controlled by the simple use of soap and water.

■ Ticks (who are immune to many insecticides anyway) are somewhat more of a challenge, but they can be removed from their host by swabbing them with rubbing alcohol and then pulling them off with tweezers when their heads come out of the skin. (Should these insects be crushed before they have surfaced, however, some of the mouth parts may remain embedded in the skin.)

■ For termites, another particularly formidable insect, there is no worthwhile substitute. It is advised that one minimize the quantity of pesticide used by just spraying the nest directly. Cryolite is the best chemical to use. Although it is highly toxic, it produces little fumes. If building a home, institute preventative measures by making sure your contractor employs termite shields, i. e., protects all wood near your foundation with sheet metal. Also, pressure-treated lumber can be used for termite prevention.

Of course, one must not lose sight of the fact that many pest problems in the home can be reduced or even eliminated by maintaining clean habits. Ants are attracted by sugar and grease. Roaches adore all kinds of food remnants, as do flies. And, needless to say, everyone knows what can be the consequences of leaving wood fragments lying around outside or inviting in a tick-infested pet.

Insect Repellents

Before leaving the subject of pest control, here's a word about insect repellents. According to the Texas Department of Health in Austin, prolonged or excessive use of such products should be avoided. Even they can be harmful, especially those containing diethyl-tolumide (deet). There have been cases in which the heavy use of this type of repellent has proven fatal to children.[7] Alternatives are vitamin B1 (thiamine) and garlic, both of which, after consumption, are excreted in perspiration and emit a scent insects don't like.

Cosmetics and Toiletries

Cosmetics and toiletries are another category to watch out for. Deodorants, perfumes, shaving cream, and the like can contribute their share of health problems. We all have our favorite brands, and we use them religiously. Unfortunately, most products of this nature, especially if scented, are predominantly a combination of synthetic chemicals. It is certainly not inconceivable to be plagued with chronic symptoms such as sinusitis, nausea, even irritability, and never equate it to the cologne that has been regularly splashed onto the skin for the last 15 years, or the talc that has been repeatedly sent into the air from an overloaded powder puff.

Nail Care Products

Nail polish and nail polish remover should be of special concern because their vapors are particularly potent before they dry. The solvents of which nail polish is composed (toluene, butyl acetate, ethyl acetate, and amyl acetate) are narcotic if inhaled in large quantities and play havoc with the nervous system.[8] The removers contain acetone or a close chemical cousin of acetone. They are highly volatile substances—and potent enough to dissolve not only nail polish but ballpoint pens, plastic jewelry, and other plastic items. The inhalation of acetone can produce headaches, fatigue, bronchial irritation, and even unconsciousness. Furthermore, these products contain formaldehyde. This is one case in which it is essential to make sure of proper ventilation before using.

Cuticle removers and softeners are also worthy of mention. They are probably the harshest cosmetic products.[9] Most contain either potassium hydroxide or sodium hydroxide. As a result, the pH level is about as much as a drain cleaner! It would be much wiser to soak your hands in a solution of warm water and mild soap.

Perfumes and Colognes

Perfumes and colognes are also composed of potent and volatile chemicals. A single scent may contain as many as 200 ingredients. Since the olfactory senses readily adapt to any odor, once applied, the wearer will soon forget it's there, but will continue to breathe its vapors. It can cling to clothes and skin, and minute amounts can be transferred to others from close contact. Even this minor displacement has been known to initiate severe symptoms in some and has been responsible for convincing a few that they had actually become allergic to their spouse or other family member. In addition, the vapors from some brands can linger in the air for long periods.

Hair Care Products

Look out for hair sprays, too. About 25% of them contain methylene chloride, sometimes disguised on the label as "aromatic hydrocarbons." Furthermore, some hair sprays may contain polyvinyl pyrrolidone (PVP). Shellac or ethyl cellulose is often included to make PVP more resistant to moisture, and plasticizers such as benzyl alcohol (derived from toluene) are also added to make the PVP more flexible. It has been noted that immediately after exposure to hair sprays, there is a measurable reduction in a person's ability to exhale.[10] Moreover, in several studies, 100% of heavy aerosol-hair-spray users were found to have precancerous cell changes. Obviously, the use of such products in this form are not a good idea. A non-aerosol pump hair spray is less of a pollutant. An even better choice is a hair-setting lotion. But keep in mind that one can still be sensitive to the emanations from either of these options.

Hair dyes can be another source of concern for those who have become chemically susceptible. Most of the ones used today are synthetic and contain additives that serve as oxidizers, color modifiers, stabilizers, and the like. One substance, aminophenol, is used in orange-red and medium brown dyes. The inhalation of this chemical may cause asthma.

Deodorants and Antiperspirants

Underarm deodorants and antiperspirant sprays are composed of such substances as aluminum chlorohydrate, alcohol, and fragrances, as well as one or more propellants. These products also contain bactericides, some of which have been found to cause health problems. One particularly harmful one, now banned, is hexachlorophene. It has been implicated in nervous system disorders and brain damage. For underarm protection, it is better to forsake spray deodorants and antiperspirants. Creams, powders, and roll-ons will at least spread a smaller quantity of contaminants. There is really no need to apply commercial deodorant or antiperspirant at all, though, when baking soda functions perfectly in this regard. Using it not only eliminates the possibility of strong chemical inhalation, but also makes things much easier on the skin.

Toothpaste and Mouthwash

Toothpaste and mouthwash can also present problems for some. They contain a number of chemicals, including formaldehyde. Baking soda is satisfactory for both purposes. A solution of boric acid can also be used for mouthwash, but it can cause a toxic reaction should a significant amount be accidentally ingested.

Shampoo

Even shampoos, which may contain formaldehyde and other strong chemicals, must be considered a menace to the sensitive. But they too can be eliminated in favor of baking soda. Using baking soda regularly is not only effective for cleaning the hair, but it also controls dandruff. Simply apply a handful onto wet hair and rub into the scalp just as you would any shampoo.

Shaving Cream

Shaving cream is another item not to be taken lightly. It contains such chemicals as phenol and ammonium hydroxide, as well as one or more of the customary propellants if in aerosol form. Plain soap and water is a reasonable substitute.

Generally, those who find themselves susceptible to any of this group of products should, if practical, discontinue use. Changing brands can help matters if switching to varieties composed of the least amount of ingredients (just the most basic ones), or those made up entirely of natural ingredients. Such products are available at many chain drugstores or department stores. Simple substitutes like baking soda and mild soap are obviously the best choices in most applications, however.

Home Improvement/Hobby Items

Also beware of home improvement and hobby items. Things like paints, paint thinners, glues (Remember the days of glue-sniffing?), stains, varnishes, and wood fillers have caused many cases of illness. Paints can prove especially troublesome.[11] They are one of the most commonly used items of this category, and some can contain as many as 300 toxic materials, including acetone, benzene, naphtha, and toluene, in addition to cadmium in the pigments.

Paint and Related Items

When painting interiors, of the two most often used products, the alkyd-base type seems to be better tolerated by most individuals than that of latex. Latex might not seem as offensive odor-wise, but it contains formaldehyde, synthetic rubber, acrylics, and preservatives such as biocide. Alkyd paints usually don't contain these things—and are better able to seal most of the vapors from suspect construction materials. (See chapter 8, *Construction Materials that Can Demolish Health*). However, they do include petroleum solvents, which can be a problem for some, at least until thoroughly dried. Epoxy paints are not recommended, because they can contaminate more heavily for long periods of time. Casein paints (pronounced ka-seen) are formulated from milk protein and, although they don't hold up as well, appear to be the best tolerated. Borax or boric acid (both usually toxic only if ingested) might have to be added, though, to retard mold.

One additional note: Use lighter colors. They contain less pigment. Off-white is a good compromise because it will include just enough pigment to tone down the brightness so as to be easy on the eyes.

At any rate, whenever embarking on such a project, regardless of the kind of paint used, plan it well. A period of "self-testing" should be performed to see which kind is most tolerable to you and your family. If a sample of paint causes any ill effects in anyone involved, that type won't be worth using. Furthermore, do it the time of year when fans can be fully blowing and windows can be left wide open. Don't use the central air system, either. This will not only distribute the vapors to every part of the house, but it will contaminate the ducts, fan, filter and other parts of the device. Also, use the brush method instead of spraying, and while work is in progress, wear a respirator mask. (The kinds available from safety supply companies are better than those generally purchased from your local hardware store.) Another idea is to "conveniently" schedule a vacation for immediately after the task is completed. That will give the paint a chance to thoroughly dry, thereby completing most of its outgassing (the release of gasses due to the curing or aging of a product).

Paint removers are worse. They contain not only benzene and methylene chloride, but chemicals such as acetone, ethyl ether, toluene, and turpentine. All are especially hazardous. Any project in which this kind of product must be employed should be reserved for the out-of-doors. Even then, use a respirator mask and make sure the wind is blowing away from you.

An extreme example, and undoubtedly one of the most tragic ever reported concerning the effects of such a product, was related by Dr. Richard Steward of the Department of Environmental Medicine, Medical College of Wisconsin. It involved a retired executive who had worked for several hours in his basement applying a paint and varnish remover to a chest of drawers. He suffered a heart attack before he could complete the job. But he survived and, undaunted, went back to the project after he had sufficiently recovered, only to experience another attack. Again, he was lucky enough to recover and eventually, never the wiser, went back to it again to suffer yet another attack—this time fatal.

Varnish is almost as bad. It, and anything else of its category that releases such penetrating vapors, is potentially harmful. Moreover, just because the label reads "nontoxic" does not mean it is not possible for some to be affected by it.

Fortunately, as far as wood stains are concerned, for those willing to give it a try, there is a satisfactory substitute. Don't laugh, but you have undoubtedly heard of or actually used the process of touching up furniture scratches by running a pecan or walnut along the wooden surface. Well, although it is more bother—and somewhat more a of challenge to keep clean after it has been applied, the ground up shells of these nuts boiled in water will perform the same function without adding harmful vapors to the air.

Glue

Glue can be a serious threat too. It contains many of the same chemicals as paint and paint removers. Watch out especially for the stronger smelling ones and bear in mind that just because a glue has dried does not mean it will no longer be bothersome. People have been known to be affected by the adhesives in items like cabinets, shelving, and paneling long after they have been constructed.

Chemical Sealers

Many caulking compounds now being manufactured are considerably safer for the air. However, there is no guarantee that some human systems cannot be affected by even the mildest of these products and it is prudent to avoid excessive inhalations.

Silicone seal, on the other hand, is an item especially worthy of mention. It releases potentially harmful vapors as it cures. Before using this or any equivalent product, take the time to read and clearly understand the warning label—and follow it to the letter. This especially applies, needless to say, if it is used often.

Other Hobby Materials

Of course, most of the above items are used in two of the most popular hobbies: woodworking and picture painting. But there are additional substances entailed in other hobbies that deserve a few words. Metalworking can involve chemicals such as sodium cyanide, sodium tribasic phosphate, and nickel sulfate, as well as solder. The former three are examples of very caustic, very toxic substances, and most solder contains 60% lead, some percentage of which is released when heated. Solder, of course, is also required when making jewelry, constructing stained glass windows, and working with electronic circuits. Extreme care must be taken when using these items. Good ventilation is an absolute must and frequent breaks are advised.

Office Supplies and Equipment

Office supplies in the home or at work, believe it or not, also rate as a perpetrator of environmental illness. Freshly mimeographed paper, carbon paper (especially NCR paper—the "carbonless" kind), typewriter correction fluid, typewriter ribbons, stencils, rubber cement, and ink (most notably the kind used for felt-tipped pens) are the most prominent examples.

These items have the capability to not only cause throat irritation, fatigue, itchy eyes, and the like, but they can also unleash severe symptoms in susceptible people. Of many reported incidents, even the child patient mentioned previously in this chapter became mentally disoriented every time she was exposed to the solvent vapors from mimeographed paper and felt-tip pens.

For that matter, even photocopiers and other similar devices, until recent years, have largely been ignored as a source of indoor air pollution. They give off vapors from the inks, dyes, and other such chemicals that they employ. Trichloroethane is one example that is used. It can produce dizziness, headaches, and possibly liver damage. Another substance included is trinitrofluorenone (TNF), which is suspected to be mutagenic. Yet another is methanol, a lung, eye, and skin irritant capable of affecting the central nervous system as well as causing liver damage. Still another is ammonia, found predominantly in the emanations of blueprint machines. In addition, these devices produce ozone, noted for its sulfur-like odor at abundant levels. Ozone is distributed in even larger concentrations if the equipment is not well maintained.

Paper Products

Even paper products make their contribution to indoor pollution. Toilet paper, tissue paper, towels, cups, plates, waxed paper, shelf paper, and even grocery sacks contain for-maldehyde, which is included in the manufacturing process because it augments the wet strength of the product. In addition, some white paper is bleached with chlorine. (Would you believe that chlorine-bleached paper products are actually banned in Sweden?) Additionally, in the case of the tinted and scented items such as toilet and tissue paper, other chemicals are added as well. If for nothing more than good measure, it would no doubt be better to maintain only the minimum supply of these goods.

School Supplies/Play Items

Don't forget about children's effects, either. Things like crayons, paste, finger paints, and clay cannot always be absolved of blame when it comes to chemically-induced ailments. They contain chemicals and produce odors too. In fact, a study by various New York consumer groups has unveiled 15 known or suspected carcinogens and mutagens lurking in children's art supplies.

Kitchen Items

Oven cleaners have already been mentioned, but other items used in the kitchen are also something that can easily be overlooked as a source of unexpected symptoms. Remember, aerosols containing cream cheese, whipped cream, and the like contain the same propellants as other aerosols. And Teflon, when heated, produces minute chemical quantities which could create a health problem.[12] Canaries, particularly sensitive creatures to air pollution, have been known to die after Teflon-coated pans were extensively used. Aluminum pans, when heated during the cooking process, can also affect some individuals. They are actually suspected of being involved in the cause of Alzheimer's Disease. Moreover, aluminum can combine easily with the food. Porcelain, stainless steel, and glass are the best receptacles to use when cooking.

Pool Chemicals

Pool chemicals are nothing to take lightly. It is certainly important to dispense and store these products properly. Chlorine, muriatic acid, stabilizers, and algae growth depressants emit strong vapors, and chlorine is known for destroying vitamin C in the human body.

By the way, keep in mind, too, that since certain regions heavily chlorinate their water supply, chlorine vapors escape into the air when water is running, producing chloroform. One should particularly beware of the shower stall because such a large volume of water is released in such a confined area—and it does so at a time when one is standing in the midst of it. Also, the hotter the water, the worse this condition will be. And splashing excessively in a tub of water releases more chlorine. Sometimes even just what stands in the area of the toilet bowl can be enough to affect some. Other chemicals, too, such as fluoride and formaldehyde can be found in tap water. It may be necessary for chemically-sensitive people to install a water purification system. In this case, the water will be much healthier for drinking purposes too.

Clothing

Even clothing is not out of the picture as far as chemicals are concerned. These products are often treated with harmful substances that serve as shrink-proofing, water-proofing, moth-proofing, and wrinkled-proofing, as well as functioning as mildew resistors, stain repellents, and static inhibiters. Formaldehyde is one of the most common and one of the worst. It is capable of causing a wide range of symptoms. (See chapter 8, *Construction Materials that Can Demolish Health*, for more information on formaldehyde.) And the unfortunate thing: There is no guarantee that it will not require many washings before the this chemical can be eliminated. If your problem is chemical-related, stay away from permanent-press, wash-and-wear, or other types of wrinkle-proofing apparel. Instead, stick to untreated all-cotton fabrics—and make them light colors or plain white since dyes (especially the darker variety) also pose a potential problem.

If in doubt about any attire, check for chemical odors that may be lingering there. In addition, if the clothing in question doesn't readily absorb water or resists wrinkling when wadded, it has most certainly been treated.

Also be warned that dry cleaning processes use a number of strong solvents, namely; alcohol, gasoline, kerosene, chloroform, carbon tetrachloride, perchloroethylene, benzene, acetone, naphtha, turpentine, and ether. If the cleaners have stored them in plastic bags, all the worse. These chemicals will be temporarily sealed in. Any apparel that has been recently dry cleaned should be temporarily hung out in the open air. For the highly sensitive, dry cleaning should be avoided.

There are many more chemical items that must be considered as possible culprits in regard to environmental illness—home permanents, skin lotions, paint thinners, etc. (See Table 3 in appendix C for a more complete list.) That is not to say that any particular product will necessarily prove to be the problem in any particular situation, but it is worth checking out.

To sum it up, avoid as many chemicals as possible if there is the slightest indication you are sensitive in this direction. Observe all instructions when chemical products have to be used. Remember that terms such as "vapor harmful" and "use only with adequate ventilation" are printed on the label for good reason. Don't neglect, either, to keep them tightly sealed, preferably in a storage area beyond the confines of your living quarters. Your lungs, as well as the rest of your body, will have a much lighter load as a result.

THE MOLD MENACE

Molds are microscopic organisms which fall under the biological heading of fungi. They reproduce by tiny seed-like spores which are released from colonies and lifted into the air. Of the many hundreds of species, some are used beneficially, as in food processing for flavoring soft drinks, alcoholic beverages, cheeses, confections, and other foods; in the manufacture of drugs such as penicillin and streptomycin; and in making silver plating possible. Out-of-doors, the mold season generally extends from late spring to fall, except in the southern regions of the country, where it can be a year-round affair. Undisturbed accumulations in spots least exposed to sunlight on tree bark, logs, leaves, soil, and the like can be a thing of beauty, often blossoming into a number of intriguing shapes and colors. Mold is also capable of growing on many other objects, including grasses, straw, and hay, in addition to crops such as wheat and oats.

Like any other contaminant, a certain amount of this matter can find its way inside. But, more importantly, some kinds of mold can proliferate inside too, and then becomes what is termed by many as "mildew." It is unsightly to say the least, and in the worst situation, if enough of the spores become airborne, they can contaminate the atmosphere to the point of causing illness. Due to continually high exposures, it is estimated that as many as 20 million Americans may have become susceptible to even minute doses.[1]

Places Where Mold is Found

Mold can take root and multiply anywhere indoors if conditions are right. It is dependent upon dampness, darkness, and the absence of ventilation for survival. Even when denied moisture, some molds can remain dormant for several years, only to revive if brought in contact with the slightest amount of water vapor.

Bathrooms and Kitchens

The bathroom is the most vulnerable place in the house because, of course, that's where most of the moisture is. And the kitchen presents its share of problems too.

The most likely places to look include the area around faucets, where there may be drips, leaky pipes, and sluggish drains. One of the best homes for mold is the bottom of cold water pipes, as is the rear base of toilets and the floor immediately surrounding them. Tile grouting is another probable spot along with damp towels, bathmats, and shower doors and curtains. Refrigerators are also a concern. Vegetable bins, drip trays, and the rubber gaskets on the doors are excellent sites for mold growth. Mold will also multiply in garbage pails.

Other Areas of the Home

As for other sectors of the house: Basements, crawl spaces, and closed-off rooms stay damp and dark. Roof leaks can create problems in attics and between walls. Leaky plumbing can be hiding under pier and beam foundations. Or the sealing around bathtubs or shower stalls can develop leaks and allow seepage into the wall or subflooring. Even awnings above open windows can be a threat. Also, improperly maintained air ducts, air cleaning filters, window air conditioners, and humidifiers accumulate a great number of mold spores. Trash compacters, too, especially if left half full, will encourage mold growth.

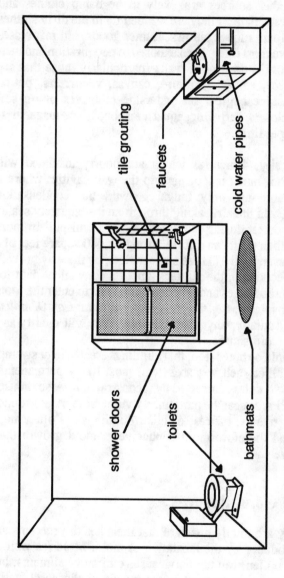

Bathroom breeding grounds for mold. The bathroom is one of the worst areas of the house for harboring mold. It can easily flourish on and around toilets, faucets, pipes, bathtubs, and anywhere else moisture is prone to accumulate.

Items That Accumulate Mold

Still other sources are likely to be damp clothes and blankets (especially if they are wadded up in the dirty clothes hamper), foam rubber pillows, leather goods, old mattresses (which have most likely been exposed to perspiration and bed-wetting), old shoes, moist walls (particularly those that are wallpapered), stuffed furniture, canvas, vaporizers, potted plants (mold can grow on underside of leaves or on soil surface), closets, carpeting, aquariums, books and magazines, and even pet litter.

Generally, structures which are poorly insulated will admit warm, moist indoor air into the wall cavities where it will promote not only fungal growth, but condensation damage. Even the air leaking through an inadequately sealed electrical outlet during the winter months will pull in more moisture than what can diffuse through 1,000 square feet of a typical plaster wall.[2] (These same outlets, by the way, as well as imperfections in the plaster, supply a convenient path for existing mold spores and other contaminants to enter the room air. See chapter 8, *Construction Materials that Can Demolish Health*.) Actually, any water-damaged area will qualify as a potential mold source.

It should be noted, too, that buildings erected over swamp-lands or other such wet areas will most likely permanently harbor mold unless corrected by reconstruction. Also, houses in wooded areas can be particularly vulnerable. And ground-floor apartments, offices, or other buildings are often more humid and therefore more conducive to mold growth than upper levels.

Health Consequences

Many are totally unaware that their health problems are being produced by the inhalation of mold. Mold (as well as dust and pollen) can not only cause respiratory ailments, but it becomes easily trapped in the mucus of the nasal cavity where it can be swallowed to cause digestive problems or find its way into the blood through the bowel to create body-wide ailments.

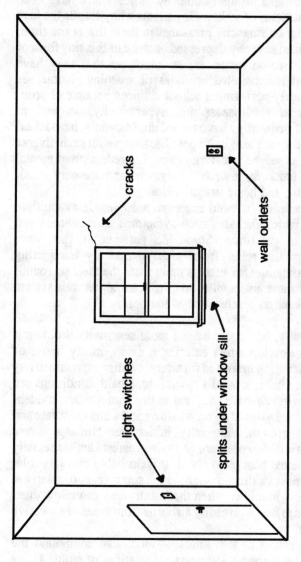

Hidden mold and its escape passages. Mold can multiply inside the walls of a house from the accumulation of moisture due to improperly installed insulation. Air flow from the attic or beneath the floor can travel along these wall cavities and push spores into the room via cracks, outlets, and the like.

Asthma, hay fever, fatigue, and depression are just some of the conditions brought about by mold. There have been reported incidents of adult male patients who, despite the lack of emotional or financial pressures in their life at the time, have remained severely depressed because of this tiny form of life, and in some instances, so much so that they have experienced uncontrolled episodes of weeping. Numerous cases of poorly-performing school children because of problems such as restlessness and hyperactivity, as well as incidents of bedwetting have turned up. Insomnia, headaches, and irritability are also quite possible along with such diverse symptoms as severe sensitivity to cold, a reduction of mental efficiency (such as impaired judgment and memory loss), constipation, and heart irregularities.

Susceptibility to mold can even masquerade as multiple sclerosis, which includes such symptoms as numbness and weakness of the limbs. Some MS patients, in fact, have obtained partial relief from their disease by eliminating airborne contaminants such as mold from the their surroundings.[3] The same also applies to other central nervous system disorders such as cerebral palsy and polio.

Generally, one could have a mold sensitivity problem if symptoms develop when entering a damp, musty house or building; if sitting in stuffed furniture; or after exposure to hay or straw in a barn, around a circus, etc. Health conditions can deteriorate in rainy weather and in the winter when condensation is prone to build up on windows and walls to initiate and perpetuate growth, especially in harsher climates where temperature differences are at their greatest, and especially when unvented heat is involved. It is probably more troublesome for most victims, though, in the summer, particularly at times of low humidity when the hot dryness can allow dust-fine spores to become readily airborne from areas of excessive accumulation.

Also, eating certain kinds of foods can compound the problem. Mushrooms and yeast are relatives of mold. Those suspecting that an airborne mold sensitivity is at the root of their problem might find it necessary to abstain from mushrooms, baked goods (such as bread and pastries), cheeses, dried fruit, smoked meats, beer, soy sauce, and the like, and shun any leftovers more than one day old, as well as eliminate the substance from their indoor atmosphere.

Discouraging Mold by Humidity Control

Obviously, in order to prevent the onset and spread of mold, the first obligation is to control humidity. Make sure the heating system is properly vented, as well as the clothes dryer, which has the habit of exhausting very large amounts of warm moisture, not to mention lint. Have air vents cleaned by a professional annually. Check rainspouts to make certain an excess amount of water isn't building up too closely to the foundation. This can make basements extra damp. If this problem exists, reroute them by using an extension or installing concrete troughs.

Check window-mounted air conditioners for standing water and supply adequate drainage if needed. Keep the filter, coils, and surrounding area clean. Also, don't forget about humidifiers. They should be cleaned and drained when not in use, and in fact, shouldn't be used at all if a mold problem exists. Keep in mind, too, that hedges planted too close to outside walls can impede airflow and retard evaporation in the lower extremities of the house. This holds especially true for walls that don't receive much sunlight. Also check for tree limbs that might be overhanging the roof.

Make certain the structure is properly insulated. A good vapor barrier is important to prevent the transfusion of moisture. (See Chapter 8, *Construction Materials that Can Demolish Health.*) Also, install plastic inserts into unused outlets, plug holes in the basement or attic where electric wires enter the wall, and make sure to patch any cracks in ceiling and wall plaster. Even examining the areas under window sills may pay off. (Note: Keep in mind that sealers can eventually dry out and shrink, and the house will inevitably shift, reopening the cracks or producing new ones. These areas must be inspected periodically.)

Wipe the surfaces of tubs and shower stalls with a towel after each use. Wash and dry bathmats regularly. Spread out used towels and stretch out shower curtains. Heavier shower curtains, incidentally, hold up best against mold. Glass doors are more of a problem, but one can control fungal growth by using a squeegee on them after bathing or showering each day. Also, hinged doors are better than sliding ones, which easily trap mold in the track. If considering the installation of a new toilet, keep in mind that there are streamlined models made from a single piece, and kinds that can be suspended from the wall, both of which present far less cleaning problems. Single-lever faucets are also a good choice for the same reason.

By rights, although the building code doesn't require one if a window is present, all bathrooms should have an exhaust fan to aid in the removal of moisture. A heat lamp should also be included. If the relative humidity is still not acceptable (at least 60% or below), a dehumidifier needs to be put into use (See chapter 5, *The Dust Dilemma*).

One must be especially responsive to conditions in the basement. Although condensation is enough of a problem, it is not the only contributor of moisture. In addition, as water accumulates in the ground around the house, it presses against basement walls, especially near the bottom where pressure is at its highest, and eventually works its way in through cracks, gaps, and joints.

Any imperfections should be patched to keep moisture outside. A good compound to employ is hydraulic cement, a quick-setting substance sold at paint and hardware stores, which expands upon contact with water. It can even be used while water is seeping through.

(Hint: To confirm the origin of a moisture problem, tightly tape a piece of aluminum foil to the basement wall when it is dry. When dampness recurs, if the side of the foil facing you is wet, the problem is condensation. But if the side that has been contacting the wall is moist, water is penetrating the wall from without.)

Other tips for keeping mold in check: Thoroughly clean or discard any article you suspect of harboring a fungal growth. If an item appears to be beyond hope, don't let sentiment get in the way. Toss it out. Even an old stuffed chair or sofa may have to go. Wash clothing often and make sure it is completely dry before putting away. Even mildew-resistant materials can be attacked if they are neglected long enough. Clothes shouldn't be crowded in closets, either. Give them room to "breathe." Be particularly cautious of cotton and kapok items. They contain an enzyme that makes mold growth more likely. In fact, apparel made of cotton or kapok which have been hung back in a closet after being worn can bring about mold within one day.

Also look under cushions, and check the bottom of furniture. Some of the heavier furnishings may be difficult to manage, but investigation could prove worthwhile. Wipe up any kitchen grease that may be lingering in out-of-the-way spots. It promotes mold growth too. Place a layer of crushed stone on the soil surface of house plants. Although it will not eliminate the mold, it will help reduce its growth. Examine all tile grouting, especially in the bathroom. If in disrepair, refill, or replace any tile that is cracked. Mold loves crevices. When renovating or building a new home, it is best to consider larger tiles. That way there will be a minimum of grout to keep clean. As it is, a small bathroom composed of 4-inch tiles can represent several hundred feet of seams.

Carpeting, especially the heavy variety, is one of the worst breeding grounds for mold. Light throw rugs are a far better substitute. At least they are easy to take up and wash. And *never* use carpeting in the bathroom, kitchen, or basement.

For that matter, never use wallpaper in any of these rooms either. It, and the paste used to secure it, is prone to collect mold. Some pastes contain fungicides to prevent this, but these can present a potential problem to the chemically susceptible. Actually, it may be a better idea to leave off wallpaper altogether. (See chapter 8, *Construction Materials that Can Demolish Health.*)

Be wary of closets. They not only present ideal conditions for mold, but they are one of the most often neglected areas when it comes to house cleaning. Some individuals may remain in a constant state of distress because of spores that waft their way into the room every time the door is opened. (Note: The problem can be reduced by allowing electric lights to remain burning in closets so equipped. Also, some people have even gone to the trouble to install exhaust fans in closets. Not a bad idea for the mold-sensitive individual.

Basically speaking, keep your home clean, well-ventilated, and well-lit. When cleaning, borax or boric acid is recommended. These substances are generally safe (unless ingested) and serve as an excellent mold-killing agent. Either can be mixed with paint and wallpaper paste, too, when they are being applied. Make sure your cleaning job is thorough.

It may mean a tough, boring job, but remember, even minute amounts of mold are potentially harmful and the least disturbance to the air can spread the spores.

In the most extreme cases, a portable HEPA filter should be employed (see Chapter 5, *The Dust Dilemma*), which will virtually eliminate any spores pulled into it.

While there is no guarantee that every trace of mold can be removed from any dwelling, following these basic guidelines will make your home more livable and you will have made great strides toward wiping out a menacing contaminant.

8

CONSTRUCTION MATERIALS THAT CAN DEMOLISH HEALTH

Stepping into a new home or building for the first time is no doubt a pleasant experience due to its newness and luster. One's eyes automatically spill over cabinets, shelves, paneling, flooring, and the other pieces of workmanship that lend to it its architectural signature. But even the closest visual appraisal will fail to disclose the fact that many of the materials used to achieve these appealing results can prove hazardous to human health.

This fact has captured the attention of many, including the California Department of Consumer Affairs which, in 1982, displayed their concern with a report stating that there is mounting evidence that offices, public buildings, and homes alike have become dangerous by contaminating the indoor air. This is not surprising considering the fact that over the past 20 years the use of chemicals in the production of construction materials has increased over five-fold.[1]

Harmful Substances Originating from Construction Materials

Formaldehyde

Although substances such as toluene and butylated hydroxytoluene (BHT) are involved in this case, formaldehyde is the chief culprit. Most think of it only as an embalming agent for lab specimens or bodies in the morgue, and indeed the morgue is where it should stay. It is an unstable hydrocarbon with a distinct pungent odor, which is used extensively in many other areas of manufacturing, and is present in such diverse products as antiperspirants, detergents, diesel fuel, paper products, fertilizers, clothing, pesticides, soft drinks, and mouthwashes. Over 7 million pounds are produced annually, half of which is consigned to construction materials. In fact, it is a $400-million-dollar-a-year industry, involved in 8% of the U. S. Gross National Product. It is often disguised on labels, too, as formol, methylene oxide, or formalin.

It seized public attention as a danger in the mid-70's when it was discovered that an alarming percentage of occupants suffered ill effects after urea formaldehyde foam insulation (UFFI) was sprayed in between the walls of their homes. One publicized account told of a New Jersey family who was out $20,000 (not counting the installation fee) to have this kind of insulation removed from their home after they began experiencing chronic attacks of headaches, nausea, irritability, nosebleeds, and respiratory problems because the formaldehyde vapors leached through the walls and into the air.

Since then (1982), UFFI has been banned, but formaldehyde continues to be present in a great number of new construction materials. It can be found in such items as plywood, chip board, gypsum board, particle board, laminated lumber, vinyl (imitation wood) panels, plaster, stucco, wallpaper, and even concrete, as well as the glue, adhesives, and the like that are used in the construction process.

Since it is so unstable, this chemical easily breaks down into a toxic gas, making it an extreme irritant. Formaldehyde is also water soluble and can readily enter the bloodstream. Excessive exposure can create eye, nose, and throat, as well as lung problems and gastrointestinal disturbances. It is also a threat to the cilia, causes the glands to manufacture excess mucus, and numbs the sense of taste and smell. More advanced reactions include headaches, wheezing, insomnia, dizziness, depression, fatigue, chest constriction, muscle and joint pains, cerebral symptoms, phlebitis (vein inflammation), heart irregularities, and even blood clotting. It is capable of aggravating a number of minor ailments such as coughs, sore throats, and colds, and triggering bouts of asthma as well as other serious illness. U. S. health authorities have yet to list it as a carcinogen, but long term exposure is a suspect in many cases of cancer. (It has produced nasal cancer in rats and mice.) Even low concentrations have been known to cause not only eye and upper respiratory irritation, but drowsiness, nausea, vomiting, and diarrhea.[2]

In fact, the National Academy of Sciences determined in 1980 that formaldehyde, even in extremely low airborne amounts, poses a serious health dilemma. One study has concluded that as many as 20% of the people in this country could have been affected by this chemical.

There have been numerous reports from individuals in every corner of the country who have suffered dire health consequences from over exposure. One very sad account told of a Waconia, Wisconsin family who underwent unbelievable agony because of particle board (which probably originated from a product lot that contained an excessive amount of the chemical). The man of the house, only in his early 30's, developed severe arthritic symptoms. Conditions got so bad, he couldn't even stand to read the newspaper because of the formaldehyde in the printing ink. Eventually, he, his wife, and 11-year-old son were so weakened, they developed sensitivities to other airborne substances such as smoke and perfume. In addition, his son actually developed brain damage.[3]

The lowest measured levels of formaldehyde in homes have been found to be 0.01 ppm (parts per million), roughly the amount present in the outside air due to automobile exhaust and other such forms of combustible pollution. Ten times this level (0.1 ppm) can be detected by most people and is the amount that has been deemed the safe limit in Canada, as well as portions of the United States and Europe. Some homes, however, have been found to exceed 10 times that level (1.0 ppm)—unquestionably an unsafe proportion.

To make matters worse, high temperature and humidity increase the rate of formaldehyde outgassing. Although these emissions diminish with time, they can still release a harmful level for as long as several years, and a tightly-closed house can perpetuate the problem by keeping them sealed within. Moreover, these vapors are readily absorbed by walls, fabrics, and other objects, which contribute greatly to bad house odor. People residing in mobile homes are at a higher risk because of the tighter construction, more compact size, and the fact that these kind of materials are more extensively used there.

Under the most adverse circumstances, there is no sufficient way to properly protect oneself. Those who suspect formaldehyde to be a health problem in their situation, short of moving or razing the place, can take some positive steps, however. Cover questionable surfaces in closets, partitions, cabinets, and the like with a low permeability paint to help seal in the vapors. Or better still, replace as much material as possible with genuine wood. Take special note of anything in bedroom areas, as well as kitchen cabinet interiors, which can even release enough vapor to contaminate food. Any offending insulation, of course, should be removed.

And believe it or not, spider plants deserve a word here. Although they cannot be depended upon heavily, air contamination research for NASA has proven that they are capable of removing small amounts of formaldehyde from the air.[4] If nothing else, formaldehyde fumes can be diluted with good ventilation.

For those who are uncertain about the content of formaldehyde in their home, there are tests that can be performed. Special badges or cannisters which absorb formaldehyde from the air can be hung inside rooms for a specified amount of time and then sent in for lab analysis. Or for more rapid results, a professional can be called in. Check with your local health department or regional EPA office.

Asbestos

Unfortunately, formaldehyde is not the only reason for concern in regard to construction materials. Asbestos is also a major threat. Everyone by now most certainly is acquainted with the danger asbestos miners and workers in asbestos plants face. Safety regulations have become sterner and strict restrictions have been placed on its usage since it was recognized as a hazardous substance by the EPA in 1972, but it can still be found in older homes and buildings. Insulation (especially in furnace ducts), tile, fire resistant coatings, troweled-on acoustic material in apartment buildings, heating duct tape, gas logs, and of course, heat shields of stoves and fireplaces can be composed of asbestos. It is also sometimes found in plaster and wallboard.

As these materials deteriorate (a process which is accelerated when humidity levels are higher), microscopic bits flake off and are released into the air. These fibers are long, thin, sharp, and very durable. Once breathed, they remain embedded in the lungs, stubbornly refusing to relinquish their configuration, initially causing irritation and ultimately tissue damage. Asbestos dust is renown for causing asbestosis, a respiratory disease characterized by a dry cough, difficulty in breathing, pain in upper chest or back, and fibrosis (an abnormal increase in the amount of fibrous tissue). Twelve or more years after these fibers have been introduced to the lungs, lung cancer can develop. And it doesn't take a large dose to threaten health. The EPA estimates that anyone exposed to more than 5 fibers per milliliter of air for any length of time stands a good chance of developing asbestosis.

There has been a special focus on children since the late 70's because of the use of asbestos insulation in schools. Much of this material has been removed, but far too many of the imposed deadlines were never met. Although the EPA has now decided that removal methods are seldom efficient enough to keep some of the fibers from escaping into the air and it would be safer to leave the material in place unless it has begun to deteriorate, as long as it is present, a possible hazard exists. The agency estimates that one-third of all schools in this country may still possess this potential asbestos problem.

Another relatively recent concern has centered around office building floor tiles containing asbestos, which have worn down from heavy foot traffic and have been discovered to be releasing fibers. In fact, a 1984 survey revealed that over 700,000 public buildings contained flaking asbestos somewhere within, including, ironically enough, 42 of the EPA's 270 facilities.[5]

Under normal conditions, the air contains less than 1 fiber per cubic centimeter. But amounts of 10 times this have turned up in homes near areas of old asbestos materials—and this is as much as 20 times the allowable level for industry! Small amounts of asbestos can be dealt with by applying a heat resistant paint or covering with duct tape to seal in the fibers. But don't take chances with anything more. It is essential that significant quantities of worn material be professionally removed.

Other Materials Which Pose Health Threats

Carpeting

As far as other materials are concerned, there are a few that warrant individual discussion. One is carpeting. If it isn't bad enough that it accumulates all kinds of health-threatening particles, it is often impregnated with dirt-repellent coatings, moth-proofing, fungicides, fire retardants, and dyes. And one of the chemicals employed for some of these purposes is—you guessed it—formaldehyde, which is the primary reason for the "new carpet" smell. (An interesting note is that these vapors have been strong enough on occasion to affect house plants in strange ways. When it is observed that stems are growing in unnatural directions, the growth rates are abnormal, or the leaf tips are curled or discolored, it may be a sign that the inside air is burdened with an unhealthy dose of chemicals.)

In addition, normal wear and tear will cause microscopic bits of the carpeting to break off and be lifted into the air by foot traffic where they are circulated through the air vents in heating and cooling systems. This can prove doubly bad in winter when they are heated because that changes their chemistry. Animal hair (such as wool) or vegetable fiber (cotton, linen, and the like) can become enough of a problem, but synthetics (such as acrylic or polyester) are capable of producing over a hundred different compounds, including sulfuric acid and hydrogen cyanide.

When purchasing carpet, it is advisable to stick with natural fibers which have not been chemically treated. (Cotton seems to be the most tolerable for the majority of individuals, although it has posed a problem for some.) Keep in mind, however, that there have been those who took the trouble to locate and install what they deemed to be environmentally safe carpet, only to discover later that they became affected by the adhesive that was used to lay it or, in some cases, the padding (which is usually foam rubber, often composed of recycled scrap). If there is any doubt, it is better to have a bare floor than to suffer. (Hardwood planks, terrazzo, or ceramic tile are good choices.) In fact, a survey revealed that 70% of the physicians who are members of the Society for Clinical Ecology preferred hardwood or terrazzo.

Wallpaper

Wallpaper is another item which can prove risky for many individuals. To begin with, wallpaper paste usually contains insecticides and fungicides. These chemicals penetrate the paper and are sometimes even incorporated into the paper itself. There are pastes without fungicides included, but using these kinds obviously encourages mold proliferation. (The best thing one can do in such cases is mix the paste with borax to inhibit mold growth.) Wallpapers made of fabric present another problem. The surface often breaks down into cellulose dust. Also, bits of fiberglass can break off of fiberglass wallpaper, vinyl wallpaper will release vinyl chloride vapors, and even ink and paints used for printed designs outgas their chemical content. In the final analysis, as with carpet, it is best not to consider wallpaper for a healthy home.

More on Insulating Materials

Insulating materials hold other hazards besides formaldehyde and asbestos. The blown-in cellulose kind is also sad news. Although it can be made from virgin wood fiber, it is more often composed of ground-up newspaper. In either case, chemicals such as ammonium sulfate have been added, and supplementing these, if the recycled paper is used, are the chemicals in the newspaper ink itself. Cellulose has a tendency to give off fine dust, and in some cases the chemicals have been known to break down in hot attics and release their vapors throughout the entire house.

Vermiculite and perlite can sometimes be a problem too. Vermiculite is made from a mica-like mineral and aluminum-iron-magnesium silicates. Perlite is composed of volcanic glass particles. Both are products are expanded under high temperature. After these types of insulation age, microscopic bits of the granules can waft their way out into the air and wind up in the lungs.

Polystyrene (known to many as Styrofoam) and polyurethane (urethane for short) may pose problems because gases such as butane, pentane, and Freon are employed during their manufacture in order to actualize the foaming process. These chemicals are trapped in the product only to gradually be released into the air.

Fiberglass, although it can eventually release glass fibers, seems to be the best choice—a fortunate situation since it represents the majority (approximately 85%) of all residential insulation. It is manufactured basically by melting the fibers, processing them to form lengths of glass wool, then spraying these lengths with a resin to bind the fibers together, and baking them to set the binder. Some have said that yellow fiberglass can be better tolerated than the pink variety, and this would especially hold true, of course, if it is isolated well from the interior.

This is one reason, by the way, why a good vapor barrier is important. If it is not adequate, mold spores, stale dust, and vapors from plastics and treated wood, as well as fibers from the insulating material will leak inside (particularly during warm weather) as wind forces its way through the attic and/ or foundation vent grills, along the wall cavity, and then through wall outlets and cracks in the plaster (including concealed ones, like those underneath window sills). By the same token, during the winter, warm interior air will be lost as it leaks through the insulation to the cooler outside wall where the moisture within it condenses, creating wet spots that will result in condensation damage and the support of mold growth.

Anyone serious about environmental chemical problems should make sure of a well-fitting vapor barrier. A good idea is to consider something other than the paper vapor barrier that is normally attached to insulation batting. It is fastened with adhesive which emits vapors that can also escape into the room. The best choice is plain fiberglass combined with a separate aluminum foil vapor barrier which should be sealed in place with foil tape. This type of vapor barrier is available through suppliers of specialty building products. When doing

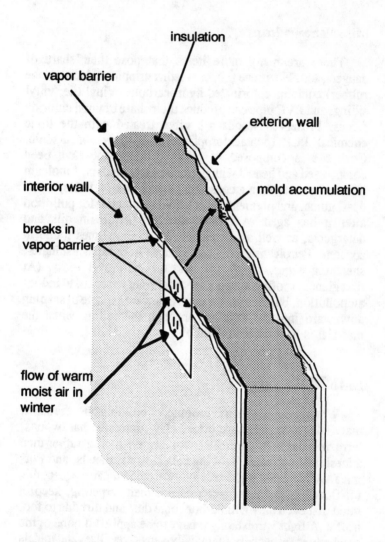

insulation

vapor barrier

exterior wall

interior wall

mold accumulation

breaks in
vapor barrier

flow of warm
moist air in
winter

Importance of an adequate vapor barrier. Walls which are
not thoroughly insulated are subject to moisture accumulation,
condensation damage, and the growth of mold due to the penetration of dampness during the winter months.

the job or having it done, make certain there is a proper seal around all wall outlets.

Miscellaneous Items

There are many more items that pose their share of dangers, too. Neoprene rubber weatherstripping (black sponge rubber) contains chlorinated hydrocarbons. Vinyl tile, vinyl siding, and PVC pipe can produce their share of vinyl chloride fumes. Preserved woods are often treated with the toxic chemical PCP (pentachlorophenol), mercury, or arsenic. Cork tile is composed of cork particles that have been compressed and heated with petrochemical binders. Linoleum is nothing more than a consolidation of cork particles, wood dust, gums, and pigments which hold a potential for pollution after it has aged. Additives such as petroleum oils and detergents, as well as formaldehyde, is sometimes found in concrete. Petroleum vapors can emanate from asphalt shingles, sheathing paper, fiberboard, and tar-and-gravel roofs. (At first glance, roofing may not seem relevant in regard to indoor air pollution, but please note that vapors are capable of seeping downward into the attic and through the ceiling when the material is heated by sunlight.)

Safe Materials

When shopping for a new home, or building one, the safest materials to consider are brick, clay, untreated hardwoods, terrazzo, glass, ceramic tile (set in cement mortar rather than adhesives), untreated building paper, stone, metals, and Formica. While untreated hardwoods are chemically acceptable for flooring, and a far better choice than carpeting, keep in mind that even they will harbor some dust and dirt due to foot traffic. A light varnish might have to be applied if none of the residents are especially chemically-sensitive. Glass and metals are both perfectly adequate for things like shelving, shelf liners, or table tops. (See table 5 in appendix D for a comparison of construction materials.)

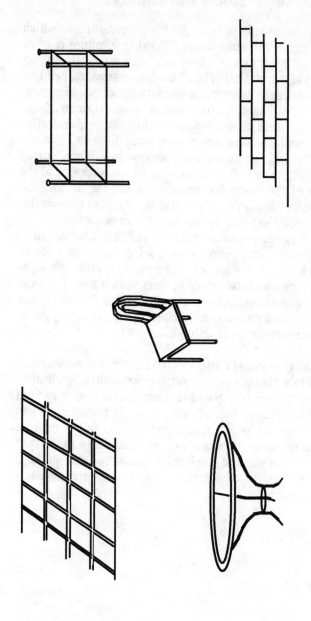

Environmentally sound construction materials. Glass, metal, terrazzo, and brick are some of the safest materials, environmentally, to be used in a home.

A Word About Home Furnishings

The same warnings apply to home furnishings, which contain many of the same things. Wood in furniture is often treated with a number of chemicals; upholstery (including synthetic and tanned leather headboards), mattresses, bedding covers, and drapes can contain formaldehyde, chemical preservatives, pesticides, sizing, and/or dyes. (Darker dyes, incidentally, seem to be the worst for many chemically-sensitive people because they contain more pigment.) In addition, upholstered furniture is likely to contain backing, a dried pasty-looking substance made of latex which is applied to the underside of the fabric for purposes of helping to hold the threads intact. Also, fiberglass drapes, which unlike insulation, is directly exposed to the inside, can become a special danger if tiny bits begin to break off and permeate the air.

Health-conscious people should select furniture made of untreated hardwoods with little or no padding. Also, wrought iron furniture and furnishings of steel with a baked enamel finish are good choices, as is wicker. Try to find untreated cotton mattresses, covers, and drapes. They usually prove to be a safe choice for most individuals.

This all may sound a bit overwhelming, but remember that the key is to make sure you are surrounded with as few health-threatening materials as possible. Only trial and error can tell you which, if any, are affecting you. If removal of any offending item at your present residence is impractical, it would be far wiser to move to a more suitable place than continue to suffer with unpleasant symptoms and perhaps facing the prospect of a catastrophic disease and an early death.

THE THING ABOUT PLASTICS

Everyone is familiar with the synthetic material known as plastic. It can be hard or soft, colorful or colorless, transparent or opaque. Plastic is probably taken more for granted than any other material because, cheaply produced and readily shaped, it is used for such a wide variety of products, from toys and transistor radios to grocery bags and automobile seat covers. (See table 6 in appendix E for a more complete list.)

Composition of Plastics

Also recognized under such headings as acetate, acrylic, polyester, polyurethane, polystyrene, cellulose, nylon, and vinyl, most people think of plastic as an inert material. But in reality, it is not. To some degree it outgasses due to the evaporation of resins (its chief constituent), plasticizers, stabilizers, antioxidants, solvents, colorants, and other such components. Plasticizers, used to maintain product softness and flexibility, can contain lead salts and derivatives of phthalic acid or of phosphoric acid; the stabilizers are made of organic compounds of tin, barium, cadmium, and zinc—all highly toxic. Solvents, which serve to make the resins fluid during production, can, as brought out in chapter 6, have their impact on health. There are over 300 possible solvents that are utilized for various plastics.[1] Even PCBs have often been used in many of these products. Furthermore, complex chemical reactions are employed in the production of plastics, which involve such substances as natural gas, coal, benzene, ammonia, ethylene, and chlorine.

Plastic Outgassing

As with formaldehyde or any other unstable matter, a warmer (especially from direct exposure to sunlight) and more moist environment increases the outgassing rate. After approximately 2 years, this process has substantially abated or stopped altogether in hard plastics (especially Formica, which is so hard it is actually almost always environmentally safe), but the softer varieties may continue outgassing indefinitely. Fortunately, one has adequate warning of this because the odor of an object is usually directly proportional to the degree of emissions. This can be easily determined by sniffing first of something like a plastic knife handle and then a plastic clothes bag.

Plastic outgassing, believe it or not, has even been known to ruin a number of space projects when the scientific instruments of rockets were rendered inoperative because of fogging brought on by emanations from certain materials that were on board.[2]

This phenomenon is evident in most new cars every morning from the appearance of that mysterious fog on the inside of the windows. It is, of course, the release of vapors from upholstery and dashboards that is responsible—the same thing which primarily gives the vehicle that pleasant "new car" smell. But while such an odor is so fondly associated with newness by some, it can be overwhelmingly troublesome to the chemically sensitive, at least for the first couple of years or so, before the process has sufficiently diminished.

In fact, automobile upholstery can emit more than 147 different organic compounds.[3] One of these substances, a highly durable plastic known as polyvinyl chloride (PVC), is extremely toxic and outgasses liberally during the first 3 months. It has been known to be responsible for chronic bronchitis, ulcers, and bone disorders. A known carcinogen and suspected mutagen, the maximum safety level of exposure to this substance has been designated by the EPA as zero.

To show just how potent these vapors can be, in 1982, more than 2 dozen prisoners died in a fire in a Biloxi, Mississippi prison when they were overcome by chemical fumes. A former mental patient was responsible when he deliberately set the padding of his cell ablaze, padding which was made of polyurethane foam—the same filler employed for most car seats, furniture cushions, and the like. In fact, research has shown that the main threat of injury or death from fires is not from being burned, but because the terrific heat steps up the outgassing process to its maximum.

Health Effects

Although no one is necessarily affected by every kind of plastic, if any at all, many have remained fatigued, achy, and otherwise ill due to plastic outgassing. Chances are if one has become susceptible to general household chemicals, there will be some weakness in this direction too. One publicized account told of a woman who advanced from a status of mildly allergic to one of semi-invalid within an 8 year period because of her habit of wrapping all her possessions (foods, shoes, clothing, etc.) in plastic. That, coupled with her bodily intolerance to office supplies, caused blurred vision, fatigue, and colon problems.[4]

Items of Concern

Some of the other common items of which one should be aware are pillow and mattress cases, plastic flooring and coverings, tablecloths, place mats, handbags, shoes (including the linings and innersoles), shower curtains, drapes, window shades, contact lenses, eyeglass frames, waterbeds, combs and brushes (both handle and bristles), and plastic furniture.

Automatic Dishwasher Interiors

Another item to watch out for is the interior of automatic dishwashers. Not that the plastic of which they are composed is all that soft, but when these devices are put to use, the high water temperature (which may be as much as 160° Fahrenheit) increases the outgassing process. That, coupled with what the soap contributes, produces anything but an odorless (and environmentally sound) kitchen. The same goes for washing machines as well as both clothes and hair dryers. Take note with your nose the next time you are engaged in any of these activities.

Lamp Shades

Plastic lamp shades are also an important consideration. They can be especially troublesome after the light bulb heats them up. There is nothing at all good about this type of shade, for instance, perched on a nightstand perhaps mere inches away from a sensitive individual. This item should be replaced with one of glass, metal, or natural fabric. The same goes for ceiling light fixture covers.

Air Purifiers

Air purifiers that should be upgrading the environment can be causing problems, too, if the casing is made of plastic instead of metal. (They can also be made of pressboard which, of course, releases formaldehyde.)

Mops

Keep in mind that some mops are made from foamed plastic resins. Plain white string mops composed of cotton twine is the best choice. Many brooms, too, are fabricated with plastic bristles. Shop around for straw.

Windows and Skylights

Plastics used in windows and skylights is another consideration. When heated by the sun, they can contribute a large share of vapors to the inside air. (Note: Some of these kinds of plastics will fail to properly shield one against ultraviolet light, which is the area of the spectrum responsible for sunburn.)

Electrical Boxes and Cover Plates

Another item, well concealed and therefore easily overlooked, is the electrical box, which is used for mounting light switches and utility plugs. Highly sensitive people may have to replace these with the metal kind. Of course, there are also the cover plates to consider. The wooden, metal, or ceramic version of these is suggested as a substitute.

Writing Pens

Something as innocuous as writing pens, both ball-point and felt-tip, can sometimes elicit symptoms. Although it might be the ink at fault, the plastic does outgas, particularly when new. It is advisable for the chemically-sensitive to use only metal ball-point pens, or pencils.

Clothing

Wash-and-wear fabrics are not out of consideration in regard to plastics. Acrylic fiber, for instance, is used for Orlon, and Dacron contains polyester fiber, one of the most toxic. Nylon, of course, is used for hosiery, light windbreakers, raincoats, and often serves as a border for blankets. In addition, plastics are used as lining in heavy winter clothing.

Electrical and Electronic Items

Plastics also come in the form of coatings. Electric wire insulation is often overlooked as a source of difficulty. Although some have blamed it on electromagnetic radiation, there are those who have been known to react adversely to a heated up TV set, an operating electric blanket, a computer video display terminal, an electric typewriter, or other such device(s). Even a heating pad, which can be largely composed of plastic anyway, is a problem source. Video display terminals and TV sets especially need to be located in well-ventilated areas because of their high operating temperature.

Teflon

Teflon, in reality a plastic, is another coating often overlooked. As mentioned in Chapter 6, Teflon-coated pans send their share of vapors into the air when heated. Some electric irons are also Teflon-coated. (And ironing board covers, incidentally, are sometimes treated with Teflon.) Even fiberglass is coated or impregnated with a plastic resin. Actually, its full name is "fiber glass laminated plastic."

Other Kitchen Items

Even plastic food containers outgas not only into the air but into food, and some studies show that as much as 50% of the chemicals can permeate store-bought distilled and mineral waters. The best kitchen materials to use are ones of ceramic and glass. Also, aluminum foil beats plastic wrap. (Beware of plastic pot and pan handles, too, especially when heated.) If purchasing bottled water, try to find glass or hard plastic containers.

Rubber

And since rubber is plastic of a sort, it must be considered as a possible source of health problems. Actually, the natural gum that is extracted from the rubber tree isn't the concern. But the synthetically-produced versions containing chemical additives designed to serve as antioxidants, stabilizers, and the like are another thing. Tires are the first item that probably comes to mind when thinking of rubber, and even they could be a factor to one who has worked for a tire company for many years. But watch out for rubber mats in the home or car, and things like sponge rubber pillows and mattresses, rubber-tiled floors, carpet padding, elastic, rubber bands, gloves, refrigerator door gaskets, typewriter pads, boots, toy balloons, false eyelashes, and even the edge of eyelash curlers. They outgas too.

In addition, it is well to say that nothing more than the body heat emanating from one sitting on sponge rubber or plastic upholstery can substantially increase the outgassing process. That makes a powerful argument for making certain not to place vinyl or other such materials too close to heating ducts.

In the final analysis, the best defense against plastic outgassing for chemically susceptible individuals is to separate themselves from as many plastic products as possible, especially the softer variety. There are many nonplastic equivalents available. If you do buy plastic, scout around for older items that have already produced most of their vapors. And don't forget that in any case of plastic outgassing in a home or other building, the problem can be reduced by adequate ventilation.

RADON REMEDIES

Undoubtedly, there a very few who have yet to realize that radon has become a serious concern in regard to the healthfulness of the indoor environment. It is severe enough to have been declared a national health hazard by the EPA. In fact, according to University of Pittsburgh physicist Dr. Bernard Cohen, "Radon in our homes gives the average American more radiation exposure than all other sources of radiation combined."

Radon is a heavy, colorless, odorless, nonflammable gas produced by the radioactive decay of radium, which is itself a product of uranium. It is chemically inert and does not combine with other chemicals or elements. Traces of radon are normally found in the atmosphere near the ground as the result of seepage from rocks and soil. On a worldwide average, approximately 6 atoms of radon emerge from every square inch of soil each second.

What Makes Radon Harmful

Surprisingly, the gas itself is not particularly harmful, but rather the subatomic particles it rapidly decays into, namely: isotopes of polonium (the most hazardous), bismuth, and lead. These radon progeny, or "daughters," as they are often called are not only radioactive, but chemically active, as well as electostatically charged. Because of this charge, they easily attach themselves to dust, particles in cigarette smoke, and other objects.[1]

The trouble occurs when these dust and smoke particles (not that tobacco smoke is not harmful enough in itself), or the unattached elements themselves, are breathed to a great enough and long enough extent. The gas itself is usually expelled, but the progeny, perhaps as much as one third of what enters in the way of attached particles and what is believed to be close to 100% of the unattached matter, lodge in the bronchial tree and lungs where they decay to a farther extent.

The high-energy substances generated by this process are alpha and beta particles, and gamma rays. All three are potentially harmful to human tissue, but the beta particles and gamma rays, because they have a higher power of penetration, distribute their radiation over a wider area, and therefore deliver less damage per unit of energy. Alpha particles, however, move more slowly and are denser, concentrating their radioactive energy on a smaller area. This disrupts the cells, causing tissue damage which, if allowed to continue, could ultimately result in cancer.

Health Consequences

These health effects are well established from studies involving uranium miners. Actually, as long ago as the 16th century, miners in Central Europe were suffering from what they called "mountain sickness," only later to be recognized as lung cancer. It was not until the 20th century, however, that such mines were discovered to contain high concentrations of radon. This prompted the tests in the 50's and 60's that ultimately established the correlation between radon and lung cancer.

In reality, the miners were fortunate in one sense. They were engaged in a significant amount of exertion. Since that created a higher rate of air exchange within the lungs, not as many radon daughters had a chance of remaining there. But in our homes, we are usually idle (either asleep or in a state of semi-rest) and the velocity of air passing through our lungs is at a minimum, giving any radon progeny that is present in the immediate atmosphere more time to stick in the tissue.

How Radon Enters Houses

As long as radon is diluted by the atmosphere, the concern is insignificant. Actually, near ground level, there is no more than about one radon atom for every 10 quintillion atoms of air (1/10,000,000,000,000,000,000,000). The problem is, it seeps into buildings and homes where it becomes trapped, and consequently accumulates. The EPA estimates that as many as 8 million homes may be harboring unacceptable quantities of radon. Other experts believe this figure to be much higher. One study concluded that radon exposure in the home could be linked to some 9,000 incidents of lung cancer annually. The EPA goes on the say that they believe 10% of all lung cancer fatalities in the United States are caused by this problem.

Cracks and Other Openings

There are a number of ways radon gas can enter a house. It usually seeps in through foundation cracks, sumps, floor drains (even those with traps to some extent because radon gas is water soluble), floor/wall joints, basement windows, and sometimes it can penetrate a sound foundation and/or solid concrete walls, since even they are somewhat porous.

Water Supplies

Radon may also occasionally sneak in via the water supply. The EPA estimates that as much as 5% of the radon exposure in the average house originates from this source, and that as many as 1,000 Americans die annually as a result of it. Although large municipal water-treatment plants are seldom at fault in this regard because the gas has ample time to decay and/or opportunity to escape from reservoirs, the smaller systems are another matter. The holding times are shorter, giving radon less of a chance to dissipate and less time to decay. In addition, the filtration process may not screen everything that reaches them.

Wells can also be a transporter if the water comes from a source which has been in contact with rock or soil containing uranium. In 1979, the EPA was shocked to discover that more

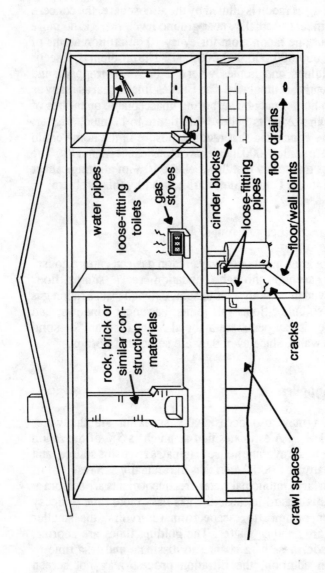

Radon entry points. There are a number of areas around the house where radon can enter, such as cracks, floor/wall joints, floor drains, and water pipes. In addition, even granite, rock, brick, or other similar construction materials can pose a hazard if they have been extracted from a radon-rich area.

water pipes

loose-fitting toilets

gas stoves

cinder blocks

loose-fitting pipes

floor drains

floor/wall joints

cracks

rock, brick or similar construction materials

crawl spaces

than 25% of groundwater samples they took throughout the country exceeded safe levels. The state of Maine has so far proven to be the worst in this regard. Tests there have turned up radon levels in well water that are 5 to 50 times the national average. This problem usually exists in wells drilled up to approximately 150 feet deep. Beyond this depth, the concentration of radon has been found to taper off. Theory has it that this is because deeper rock is less porous and because water in deeper wells remains there for several days before it is pumped to the surface, giving any radon that is present more time to decay.

When water that has absorbed radon is used (especially in spray devices such as showers or when heated as in cooking), the gas is released into the air to accumulate inside a bathroom or kitchen. Actually, there's plenty of opportunity for radon-contaminated water to release its poison. It is surprising the amount of water that is used during normal household activities. Not only are showers, baths, and cooking involved, but also house cleaning, brushing teeth, washing hands, shaving, doing laundry, using the dishwasher, and flushing the toilet.

One might be tempted to say that using water in this condition for drinking purposes isn't particularly healthy either. And this is probably true. Although the digestive system is considerably more rugged than the respiratory tract and the radon daughters generally don't have the punch to get through the stomach, residual amounts can find their way to the liver, kidney, or other organs and possibly be the cause of health problems. There has not yet been much experimentation in this direction, however, and a lot of questions are still unanswered.

Construction Materials

Also, there have been some cases in which radon problems resulted because of the gas being released from granite, brick, concrete, clay, or other such material used to build a structure—or even the rocks contained in a solar heating system. Granite, although not used much in home construction, can be found in some apartments, office buildings, and college dormitories, and is one of the worst offenders because its uranium content is so high. But any material can be problematic if extracted from a radon-rich area.

This implies that rock collectors should be wary. Some specimens may be composed of minerals such as carnotite, titanite, pitchblende, or zircon that contain radon-emitting uranium.

Natural Gas

Furthermore, as previously mentioned, radon can also be found mixed in with natural gas. The EPA estimates that on an annual basis, over 140 million Americans are exposed to at least a notable quantity of radon which has seeped in through unvented ranges and space heaters.

Radon Rich Areas

Concerned individuals should check with their state or local Environmental Protection Department or the geology department of a nearby university to determine if they are living in an area with a potentially high radon level. Certain locations in Maine, Illinois, California, Idaho, Montana, Oklahoma, eastern Pennsylvania, eastern Kansas, Maryland, New Jersey, New York, western Missouri, Wisconsin, and central Texas have long been established as particularly serious trouble spots. The town of Clinton, New Jersey alone contains a subdivision of houses built on a ridge adjacent to a limestone formation that is radioactive. And an extensive deposit of uranium-laden granite called the Reading Prong, extending through portions of New Jersey, New York, and Pennsylvania, and involving approximately 100,000 homes, has proven to be one of the worst areas.

Also on the alert list are regions that were former mining areas where uranium tailings exist such as Florida, Tennessee, and Colorado. Some of these tailings were once used not only for building materials, but for landfill as well. Grand Junction, Colorado has been particularly plagued with this problem.

Additionally, the long unsuspected state of Iowa has been recently implicated. In fact, the EPA was astounded to learn that a whopping 71% of the homes tested in early 1989, exceeded safe levels.

Tests completed by the EPA in the fall of 1989 have even turned up astonishingly high radon levels in other states, too—namely Georgia, Ohio, Vermont, New Mexico, West Virginia, and Alaska.

For those living in these or other potentially troublesome parts of the country, it is strongly advisable that they test their home for possible contamination. In fact, the EPA has recommended that everyone conduct a radon test, because the problem is turning up more and more outside designated trouble spots. Be especially wary if you have a basement, a drilled well, or any crawl space that is unvented.

Radon Testing

Professional

There are a number of testing options. Some involve the use of a variety of very expensive laboratory instruments used by commercial establishments. Most give readings in working levels (WL), which is the unit of measure for radon decay products. For various fees, a professional can be called in to take measurements with one of these machines. Readings may not always reflect an accurate average, though, because such mechanisms are not only intended for short-term testing, but in some cases are sensitive to weather conditions or excessive dust. They do, however, deliver quick results and can help give one a general idea of the condition of his or her inside air. Check "Environmental Services" and "Radon Detection, Measurement, and Correction" in the yellow pages.

Charcoal Adsorption

There are also two types of devices that can be purchased and placed in the home for measuring inside levels of radon, both perfectly satisfactory for the job and all one really needs to establish whether or not a health hazard exists. They come in the form of inexpensive kits. One is called the charcoal adsorption test; the other is the alpha track (or track-etch detector) test.

Charcoal adsorption, the least expensive (from $10.00 to $40.00), is merely a metal, plastic, or paper container of activated charcoal granules which collects radon gas. Capable of detecting a level as low as 0.03 pCi/L (or pico Curies per liter, the unit used to measure the quantity of gas in the air), it is placed in a strategic spot a few feet off the floor for several days and then mailed in for analysis. It is quick, reasonably accurate, and easy the use. But it is not to be mounted in humid places like bathrooms and kitchens or in areas of high heat such as near stoves. It is moisture-sensitive as well as heat-sensitive and will not render reliable results if so affected.

Alpha Track

Originally developed for the Apollo space program, the somewhat more costly alpha track test ($25.00 to $50.00) consists of a small plastic cup surrounded by a strip of photographic paper. This paper, which must be unspooled and hung several feet from the floor, is sensitive to alpha particles and records any emissions reaching it. These readings are also expressed in pico Curies. It is designed for long-term testing (one month to one year) and is not affected by humidity or temperature. In some cases, especially with lower radon concentrations, it may not always be as accurate as the charcoal adsorption test, but nonetheless it is usually dependable and just as easy to use. The most accurate average, of course, will be achieved by conducting a year-long test. Like the charcoal adsorption test, it must be sent in to a lab to obtain results.

For a current listing of approved companies that offer charcoal adsorption and/or alpha track testing services, write the EPA Office of Radiation Programs. (See Appendix G.) Radon self-test kits are also now available at local supermarkets and hardware stores.

(Note: Some may be tempted to use a Geiger counter to detect the presence of radon, but it is useless in this case because it is not sensitive to alpha particles.)

How and Where to Test

No matter what test is used, it is advisable to give every room attention. The basement ranks as the most important area since, being underground, it is the closest to any source of radon, and because it represents the largest surface area exposed to the surrounding soil. Moreover, the pressure is less there than that which is in the ground, making it the path of least resistance. (Houses with basements average approximately 30% higher radon levels than houses without basements.) For two-story or multi-story dwellings without basements, the first floor, especially bedrooms, should have priority. In other words, the level of the structure closest to the source of radon will have the highest concentrations. (However, that does not mean there will not be a radon problem at higher elevations. Elevator shafts, stairwells, and air ducts make an excellent avenue for ascending gas, so high-rise apartment residents beware.)

Be sure, too, that ventilation is kept to a minimum during the testing period so that the highest possible reading can be obtained. That way, you will have established the worst-case scenario and will be better able to determine what measures must be taken.

Just How Much is an Unsafe Level?

Not everyone agrees on just what a safe level of radon is. We do know that lifetime exposure to 200 pCi/L will boost the odds of dying from lung cancer to more than 75 times the nonsmoker risk, that 100 pCi/L is equal to enduring 10,000 chest x-rays per year, and that even 10 pCi/L carries a health risk roughly equivalent to smoking a pack of cigarettes a day.[2] The general consensus is that anything below 4 pCi/L is not a significant reason for concern. (It is said that persons living their entire life in a house containing 4 pCi/L may have little more than a 1% chance of developing lung cancer because of radon.) But this is by no means a magic number. There are those who maintain that any degree of exposure is harmful. Obviously, the lower even the lowest level can be brought down to, the safer the air will be.

WL	0.001	0.005	0.01	0.02	0.05	0.1	0.2	0.5	1
pCi/L	0.2	1	2	4	10	20	40	100	200
times average outdoor level	average	5	10	20	50	100	200	500	1000
times average indoor level		average	2	4	10	20	40	100	200
comparable risks (Based on lifetime exposure)		minimal risk	200 chest X-rays a year	3 times non-smoker risk	1-pack-a-day smoker or 5 times smoker risk	2-pack-a-day smoker or 2000 chest X-rays a year	30 times non-smoker risk	10,000 chest X-rays a year or 4-pack-a-day smoker	More than 70 times non-smoker risk

Radon exposure and lung cancer. The above information displays the lung cancer risks posed by radon exposure as compared to other causes of the illness. Figures are approximate. Source: Adapted from Lafavore, Michael, *Radon: The Invisible Threat* (Emmaus, PA., Rodale Press, 1987.)

Factors That Vary Readings

Don't be fooled, either. The extent of this condition can vary wildly from house to house. Levels can range from as little as 3% of the 4 pCi/L allowable standard to as much as 6 times this amount or more. The integrity of the structure, the degree of ventilation within that structure, the pressure level of the structure, the chemical composition of the surrounding soil, the permeability of that soil, and the absence or presence of ground fissures are factors to be considered.

According to some experts, soil permeability may just be the most important factor.[3] One test conducted in an area of New Jersey showed that as many as 50% of the homes constructed over loose soil had dangerously high radon levels, while nearby houses built on more compact ground were normal. A similar study in Tennessee supported these findings.

The intensity of an existing problem can also vary from season to season or adjust itself when there is a change in the weather. Rain, a freezing ground, or a rise in barometric pressure curb the escape of radon. Wind or a warm, rising air mass, on the other hand, tends to have the opposite effect. As it turns out, for reasons as yet unclear, wintertime testing seems to be more consistent than summertime testing, and on the average, indoor radon concentrations are about 60% higher in winter. Perhaps it is due at least in part because of the fact that for many, summer is a time for increased ventilation which, of course, would result in diluted levels. Another reason is most likely attributable to the rising warm air from heating systems, which creates a slight vacuum in the lower extremities of the house that in turn lessens pressure and encourages radon entry. No firm facts on this have been established, however.

If you discover that there is a radon problem, immediate action is in order. There are a number of measures which can help to remedy the situation.

Ways to Stop Radon

Sealing Cracks and Other Openings

First, seal all potential routes of entry with caulking or cement. Look for cracks in the foundation, floors, and basement walls, as well as openings, such as where pipes come out of walls (not only those employed for plumbing, but those used for electric utilities and telephone lines), or joints between walls, floors, and fireplaces. The slightest fracture may be adequate for an invitation to radon. Even the area around the base of toilets should be checked. Sumps should be caulked in a tightly-fitting cover and vented to the outside with adequate piping. Don't neglect enclosed porches either.

If the suspected culprit is stone, brick, or concrete on an interior wall, it should be sealed on the inside with cement mortar and a non-porous air barrier. Several coats of epoxy paint will usually suffice, but one must take into regard anyone in the family who is chemically-sensitive. Don't forget, too, that paint (as well as caulking and other sealers) will eventually deteriorate and must be renewed or a radon problem can resurface.

The focus should be on basement areas, not only because of the reasons previously stated, but because the walls are often constructed of cinder blocks, which are porous, as well as hollow on the inside. (The top of block walls, if accessible, as well as the interior side should be sealed off for maximum protection. If the block tops are not accessible because they are covered by the sill plate, seal in the seam between the sill and the blocks.) Also to be reckoned with in some basements is a French drain, an intentional gap up to 2 inches wide which is left between the walls and the slab floors for purposes of drainage in the event condensation forms on the walls. (More on this later.) Keep in mind, too, that basements with water leaks mean potential radon leaks. Not only is this a sign of poor structural soundness, but nearby radon in the ground will merely dissolve in the water and be carried in.

Also note that homes built on pier-and-beam foundations are more likely to possess a radon problem than those situated on solid slab. Radon gas can be drawn upward with warm air currents. It can also slip in via improperly sealed air conditioning and heating ducts that have been routed through the crawl space. Make certain this area is adequately ventilated. Also, the installation of a vapor barrier under the floor will help.

Increasing Ventilation

Increase ventilation in the house, too. In less than severe cases, it can work wonders. The more rapidly the air is exchanged, the less build-up of radon gas there will be. Actually, these accumulations are pretty well inversely proportional to the ventilation rate, that is, doubling the ventilation rate will half the radon level. Just opening windows is remarkably effective. It is estimated that the average reduction by employing this simple method alone is near 90%. The use of fans to pull in fresh outside air helps even more. Not only does this increase ventilation, but it also raises the inside pressure, which reduces the entry of radon by reversing the air flow.

But as effective as this these tactics may be, they take on a different light in the winter. In this case, an air-to-air heat exchanger, or heat-recovery ventilator, can more comfortably help keep a home aired out than open windows and fans. It is a device in which stale indoor air and fresh outdoor air is exchanged by passing through ducts with a common wall that is very thin and easily capable of transferring heat. There are window or whole-house units available. The whole-house version, of course, is more efficient, but some ductwork will have to be added. The average radon reduction rate with these devices is over 95%.

Sub-Slab Ventilation System

If a basement is involved, blowing in air from the upper level may be sufficient to equalize the pressure. If that fails, there is another option: the installation of a sub-slab ventilation system.[4] It is a contrivance consisting of an exhaust fan and pipe (or a network of pipes) that is used to pull contaminated air from under the basement floor and vent it outside. The pipes must extend below the floor and be routed up along the walls to a common point and then outside to the fan. It is a well-proven method, but must be accompanied by a layer of gravel under the slab in order for the fan to generate enough suction.

Pipe-in-Wall Method

Even sub-slab ventilation may not be adequate if radon is sneaking in through cinder block walls. In this case, it will be necessary to employ a variation of that method, known as the pipe-in-wall approach, i. e. piping must be installed into the hollow centers of some of the blocks and routed outside to an exhaust fan.

Baseboard Duct System

Another variation of this technique will turn a French drain from a determined foe into a godsend. Known as a baseboard duct system, it involves the air-tight installation of a sheet metal baseboard around the entire length of the drain and the connection of that structure to an exhaust system.

Drain-Tile Suction

Still another application, drain-tile suction, entails the use of drain tiles, an assemblage of perforated pipes surrounding the foundation of some houses for the purpose of diverting water away from basement walls. As long as these pipes form a complete enclosure, it is possible to pull radon from the soil under the house by attaching them to a pipe/fan exhaust system. In some cases, the drain tiles will be connected to a sump, in which case the exhaust system will have to be routed from it.

For detailed information on this and the previously mentioned methods, see *Radon: The Invisible Threat* by Michael Lafavore, *Radon: Risk and Remedy* by W. H. Freeman and Company, and *Radon: A Home Owners Guide to Detection and Control* by Bernard Cohen.

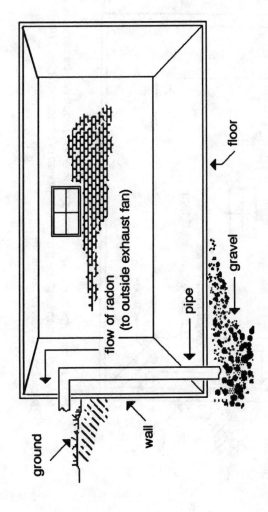

ground

wall

flow of radon
(to outside exhaust fan)

pipe

gravel

floor

Sub-slab ventilation system. *Although this method is often effective in reducing radon concentrations, there is no guarantee it, or it alone, will prove to be the answer.*

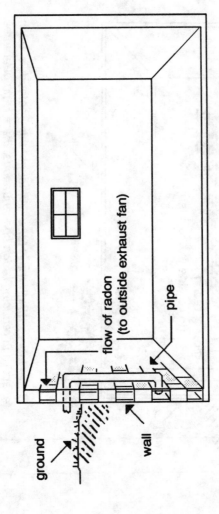

Pipe-in-wall method. This method of diverting radon can be particularly useful when basement walls composed of cinder block have been allowing the gas to penetrate.

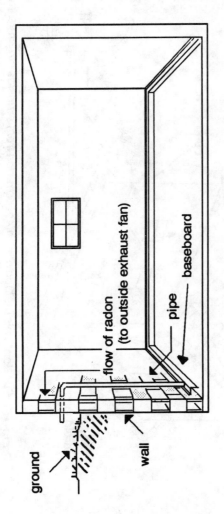

Baseboard duct system. *This techique, employed in basements with a French drain, involves the use of a sheet metal baseboard connected to an exhaust pipe.*

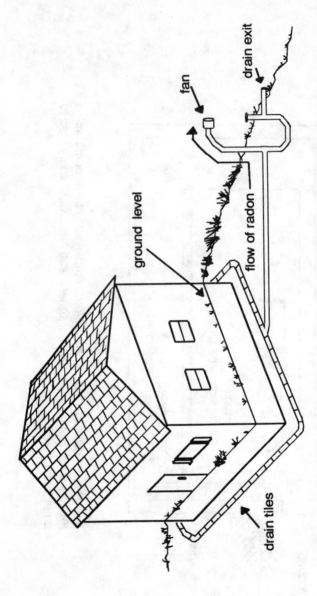

Drain-Tile Suction. This method of eliminating radon from an inside structure involves the use of drain tiles, which are connected to a pipe/fan exhaust system. The exhaust system can be routed through a sump if drain tiles are attached to the sump.

In the above cases, however, it is wise to seek a competent professional if you plan to institute such installations, as they can be tricky. Every inch must be thoroughly sealed in order to achieve adequate suction. Furthermore, since each house is structurally unique, a combination of several methods may have to be employed for satisfactory results.

Filtration

Filtration has also been used to ease radon problems, but some experts claim it is considerably more effective for water than for air because smaller dust particles, the ones that escape filtration (and are more likely to enter the lungs) will just as easily be attracted to radon daughters. And it does nothing to screen the unattached progeny themselves. In fact, this process may actually make matters worse because since it reduces the quantity of dust in the air, it leaves less targets for the progeny and increases the amount of unattached progeny in the air. Moreover, even a HEPA filter's efficiency is dependent upon the amount of particulate matter it actually pulls in. What misses the filter is still free to be inhaled. The best answer as far as the air is concerned is to stop radon before it enters the premises.

A granular activated carbon (GAC) adsorption unit, which is a tank containing a bed of activated carbon, is the best device to use for water. There is, however, no guarantee that an appropriate level can be screened even from water if the situation is severe.

The safest policy is not to play games with radon. It is something that should be stopped as soon as possible. Although a few more months of procrastination won't make much difference compared to a lifetime of exposure, it won't exactly help matters any. And if a radon problem goes unchecked, irreparable body damage can eventually result. Under extreme circumstances, professional assistance is recommended. Seek advice from the EPA; local, county, or state health authorities; or, in some cases, even your local utility company.

THE POLLEN PROBLEM

Anyone who has picked up a dandelion has most assuredly noticed the yellow "powder" which so liberally comes off on fingers and hands. This, of course, is pollen—the fertilizing element which, largely carried by wind or insects, is responsible for perpetuating plant reproduction. There are many kinds of pollen, each produced from its particular species of plants, including those of grasses, trees, and weeds.

Unfortunately, it plays one of the largest roles as an instigator of respiratory ailments. According to American Academy of Allergy estimates, 35 million Americans are troubled by pollen. The most common reaction is pollinosis, better known as hay fever. Other problems that have been known to occur are asthma and depression. Also, it is not uncommon for people who are sensitive to mold to be equally affected by pollen, nor is it unusual for persons to react differently to different kinds of pollen. In fact, just one microspore of pollen can trouble a highly sensitive individual.

The pollen season can be miserably long in warmer climates where the growing period is of a lengthier duration. Moreover, those susceptible to more than one kind of pollen, must face an even greater burden. With few exceptions, trees contribute the most in springtime, grass presents the heaviest problem during early summer, and weeds are the major culprit by the fall. Virtually all plant life is capable of generating pollen until it has gone to seed.

Which Sources are More Likely to Create Health Problems?

Weeds are the worst offenders—particularly ragweed, because it is so prolific. One of its favorite habitats is freshly disturbed ground (a special invitation being uprooted dirt at construction sites), but it grows abundantly in most any soil. It is such a hearty breed, that ragweed seeds have been known to germinate 40 years after production. Although relatively inconspicuous, it can be recognized by its coarse hairy stems, divided leaves (with 3 to 5 lobes), and long spikes of tiny greenish-yellow flowers. There are hundreds of varieties. Allergists in the U. S. estimate that in excess of 250,000 tons of ragweed pollen are picked up by the atmosphere annually. One plant is capable of producing as many as one trillion pollen grains!

Grasses are the second worst offenders. Of the some 4,000 species, Bermuda grass contains the most allergenic pollen. In the mid-south and Gulf states, it is a usual part of the landscape.

As a general rule, flower pollen is less likely to trigger symptoms because, not only is a smaller quantity of it produced, but it is larger, heavier, and excessively sticky, and therefore not as easily transported by the wind. Its transfer is usually dependent upon insects, which are attracted by the colorful petals and sweet fragrances of the flowers that produce it. But tree, grass, and weed pollen is readily carried aloft (sometimes thousands of miles) because of its smaller size and the fact that it is not particularly sticky. There is nothing to prevent it from landing on flowers, either, so take heed if bringing roses, iris, or the like into the house.

House plants, by the way, rarely manufacture pollen. It's the mold that grows on their leaves and soil surfaces that can pose a problem.

Pollen is at its worst on hot, dry, windy days, and less troublesome when it is cool, cloudy, and windless. Conditions are also more favorable every time it rains because water washes it out of the atmosphere, and when the humidity is too high, because the pollen is not released then. In addition, levels are generally highest in late evening and early morning,

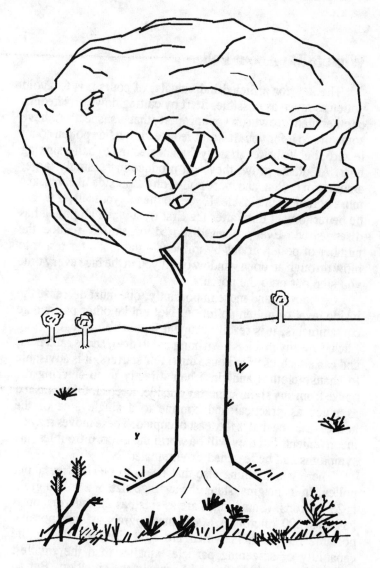

Sources of pollen. *Millions of Americans have developed allergies to pollen. Trees contribute most of the pollen in the spring, grass presents the heaviest burden in the summer, and weeds predominate in the fall.*

so keep this in mind when airing out the house or going for a walk.

What to Do About Pollen

The key for afflicted individuals, of course, is to avoid pollen as much as possible. Start by cutting down any hedges or other plant growth on your property that is suspect—or have someone do it for you. It won't reduce the outside pollen count to zero, but its better strategy than neglecting to bail a sinking boat. As plants grow, they will produce increasingly larger amounts of pollen, and more pollen on the outside likely means more pollen on the inside. If you do the job yourself, it would be better to wait until after the first frost when all pollen has disappeared. Every plant disposed of should reduce the number of pollen grains by millions—and that's far fewer to blow through an open window or hit you in the face every time you step out onto the porch.

Obviously, and more importantly, one must decrease the inside level of pollen pollution. Not unlike other particulate contaminants, it is readily brought indoors on hair, clothing (including any that has been hung out to dry), shoes, and pets, and can attach itself to house dust. For starters, it is advisable to change clothes and rinse hair directly upon entering the house from any significant stay outside. Keep clothing washed as much as practical and confine to a single area of the house—the one that is the least occupied. These moves may be inconvenient, but they will be worth the effort if troublesome symptoms can be lessened or eliminated.

There is something else that needs to be clarified about pollen. In its original state, the average size of a pollen grain is 25 microns. (Some kinds are as small as 2.5 microns or as large as 200 microns.) Many have purchased inexpensive portable air purifiers which are advertised as having the capability of screening particles smaller than the smallest pollen size, thinking this will remedy the situation. But an often overlooked fact is that pollen breaks down into fragments which are considerably smaller. Often, afflicted individuals can stay closed up tightly indoors only to have symptoms persist because very minute particles escape not only an air purifier, but the less efficient air conditioner filter as well, and perpetually circulate through every part of the house. In fact, studies have even found bits of pollen inside homes months after all plants have frosted over.

If facing an indoor pollen problem, the most thorough means of removal rests with HEPA filters and water-trap or central vacuuming systems. (See chapter 5, *The Dust Dilemma* for the details on these devices.)

It must be mentioned that some resort to moving to escape from the kind(s) of pollen native to their region. But this is only advisable as a last resort. Many find to their dismay that they can develop sensitivities to other kinds of pollen. For instance, a person could locate an area free of the ragweed they have been highly allergic to only to eventually inherit a problem with Russian thistle. It is difficult to know if this is going to happen until one has lived in a new place for a considerable period of time.

For the record, the most dense ragweed concentration is in the midwest from west central South Dakota to Ohio, northern Pennsylvania, central New York, most of the east coast, eastern Montana, and western Texas. To avoid ragweed, try the desert regions; southern Florida (including the keys); forest areas of the Rockies, New Hampshire, and northern Maine and Michigan. Generally, the best places to escape pollen are the seashores, where the prevailing winds move in from the ocean, and high mountainous areas, where the cool, thin air has less ability to transport it. (See appendix F for a breakdown of some common allergy-producing pollen sources and general guidelines about the locations and time of year that they are active.)

GUIDELINES CONCERNING GASOLINE

The use of chemicals in regard to a healthy indoor environment has already been discussed, but gasoline (and the combustibles it produces) deserves separate attention.

Although it is sometimes used in the home, automobile repair garages, or other such locations for cleaning purposes, its primary role is, of course, that of a fuel for the internal-combustion engine. Everyone recognizes one of the by-products of this combustion as carbon monoxide. But nitrogen oxides, also a by-product, and bad enough in themselves, combine with the benzo-a-pyrene and the hundreds of other hydrocarbons of which the gasoline is composed to produce secondary pollutants such as nitrogen dioxide, ozone, peroxyacyl, formaldehyde, and acrolein. To make matters worse, some gasoline still contains lead. These contaminants are of large concern. It is well established that they contribute heavily to outdoor pollution.

How Exhaust Fumes Enter Vehicles

But there is more to consider than meets the eye. The fumes, all too often, build up inside operating vehicles and prove particularly potent to passengers occupying such confined spaces. A fact unrecognized by many is that people traveling in back seats with windows open—even a crack—are exposed to a dose of their own vehicle's exhaust fumes because the vehicle's motion creates a vacuum behind itself which will capture some of these contaminants and suck them inside.[1] Rear windows of station wagons can be especially bad. Numerous cases of car sickness may really be nothing more than a set of symptoms that have been brought on by this phenomena. The first duty of those prone to illness brought on by this is to keep all rear windows closed. In fact, even front windows, to some measure, can pose the same problem.

Heavy traffic can pose problems for even front seat occupants. Exhaust from other vehicles can enter through windows and air vents. It is advisable under these conditions to keep all windows up. Don't use any outside air intakes, often labeled "fresh air" or "mix" on the dash, because it will do nothing more than bring in "fresh" air from the road, containing not only exhaust fumes, but sometimes a host of other unexpected surprises such as emanations from freshly tarred streets, factories, sewers, or even pesticide sprayings, as well as fumes from the engine. The same thing applies to the heater because it also pulls in these outside contaminants. Dress appropriately and don't use it unless absolutely necessary. In summer, it is often difficult to escape running the air conditioner, but even many factory installed units will invite in a certain percentage of outside air. If this is so in your case, it is a good idea to have one of the re-circulating variety installed. In the event of severe situations, there are portable air purifiers especially designed for automobiles which, when plugged into the cigarette lighter, can filter some of the pollution that has entered. Examine the yellow pages under "filters."

Those sensitive to automobile exhaust who have control over the times they travel should do so in the mornings when the air is fresher or the middle of the night when there are less people on the road. This especially holds true in the summertime when the hot afternoon air is usually packed with a higher measure of pollution. When there are no cars nearby, the windows should remain open so that the interior air is renewed. Use judgment in how you drive by simple observation during daylight hours. Be alert to which way the wind is blowing by watching which direction the smoke from factories, diesel trucks, or the like is moving, or perhaps taking note of tree movement. Then adjust your driving accordingly by remaining on the desirable side of vehicles when possible. Also stay a reasonable distance behind the vehicle ahead of you. And avoid tunnels, which are notorious for trapping fumes.

Of course, vehicles in disrepair, such as those with defective mufflers or ones that are poorly tuned, must be attended to. As might be suspected, the more smoke that is visible from an exhaust system, the more inefficient the engine is operating and, therefore, the more combustibles there are released into the air. Other more obscure problems can also crop up. The young athlete mentioned in Chapter 3 that had been affected by natural gas suffered with a recurrence of his symptoms when a crack in the rubber at the base of the floor-mounted stick shift of his sports car allowed vapors from the heated-up engine to seep inside the snug, two-seat compartment.

Diesel fuel emissions are more health-threatening, especially to those who have developed a sensitivity to petroleum derivatives.[2] This fuel is composed of heavier hydrocarbons and it vaporizes at a much higher temperature (some 3500° Fahrenheit, as opposed to no more than 700° Fahrenheit for standard gasoline). In fact, tests have proven that petroleum products boiling at over 700° are carcinogenic to mice, rabbits, and monkeys. Passengers on busses have been adversely affected when sitting in the rear, or even when occupying the front if other busses are around.

How Exhaust Fumes Can Affect Health

A whole slew of symptoms can be initiated and perpetuated by gasoline exhaust vapor. Headaches, fatigue, and confusion head the list. Other complaints include depression, nausea, dizziness, feelings of suffocation, and irregular heartbeats. The effects can be subtle too. Since it often plays havoc with the nervous system, it can interfere with coordination and slow a driver's reaction time. "Driver's hypnosis," the sneaking grip of sleepiness and fatigue, may often be attributed to vehicle exhaust. There is every reason to suspect, too, that it has been implicated in accidents where "irritable" and "impatient" drivers were involved.

Those who are heavily exposed are particularly vulnerable. A number of truck drivers, cab drivers, route salesman, automobile mechanics, toll booth employees, and the like have eventually succumbed. Gas station attendants, although a dying breed, have another problem in that they deal with a larger than normal amount of raw fumes. They, and the "self-service" community of America, should avoid breathing the vapors while filling up, and stop the first time the pump clicks off.

Actually, the raw fumes as well as the burning emissions can be addictive to the body, too. Many have suffered a round of depression shortly after filling their tank (as well as engaging in activities such as riding in the rear of a car or bus) because they had unknowingly gotten "high".[3] Other symptoms as a result of over-exposure to the unspent vapors include flushing, staggering, slurred speech, and confusion. It is always wise to avoid fumes of this nature even if—or especially if—they seem pleasant to smell.

Gasoline and the Indoor Environment

Although there is justifiable concern regarding the health aspects surrounding this chemical when dealing with motor vehicles, both from within and without, gasoline poses an equal worry in its relationship with the environment of homes and other structures.

Of course, gasoline should never be used or stored in the home. That's easy enough to control, but even if such advice is strictly heeded, too often its vapors are, unknowing to many, invited inside anyway. Storing filled containers in attached garages will not guarantee that the living quarters will be isolated from gasoline vapors.[4] Some amount can leak from the container, fill the area, and not only sneak into any heating/ air conditioning ventilation connections installed there, but penetrate the wall, seep through the extremities of a doorway, or escape via the attic and make its way back into the rooms of the house from the ceiling, especially in the area around the light fixtures. This goes double if you fill up the lawn mower from inside the garage. A separate shed is a worthwhile investment to store gasoline (and other chemicals), as well as lawn mowers, powered tillers, and the like, which house fuel not only in their gas tanks, but in the lines and carburetor as well.

Carbon Monoxide

Naturally, everyone is aware of the hazard of running a car for any extended length inside a closed, occupied garage. To state it simply, the atmosphere of an unventilated single car garage will become lethal within approximately 5 minutes if a car motor is left running. Although one should generally be repelled by the odor of the emissions, the component of most concern, carbon monoxide (representing as much as 7% of the fumes), is colorless, odorless, and tasteless, and therefore cannot itself be detected. It is readily absorbed by the bloodstream (possessing 200 times more affinity for hemoglobin than that of oxygen), and severe symptoms or even death usually occur before one is able to realize what is happening.

For the record, initial symptoms of carbon monoxide poisoning include tightness across the forehead and sometimes a slight headache. As exposure time increases, a headache will definitely begin to develop and there will be throbbing in the temples. A severe headache will follow, along with weakness, dizziness, dimness of vision, shortness of breath, nausea, vomiting, and finally collapse.

But even if the garage is unoccupied, the routine of leaving the family car running has sometimes produced disastrous results in an unexpected way. There have been reports of accidental deaths in bedrooms adjoining a garage where an automobile had been idling.[5] And just remember that lawn mowers and other similar devices can be just as bad. It is an extremely poor practice to service or operate any gasoline-powered apparatus inside the garage.

Other Emissions

But that's not the half of it. Cars should remain outside to cool off before being put into the garage—especially if one is dealing with an attached garage built directly below a second

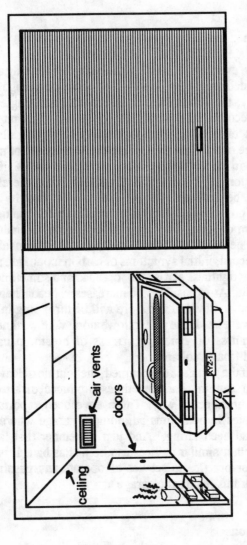

How gasoline vapors can enter a house. Although it is obvious that a vehicle idling inside a closed garage can result in carbon monoxide poisoning, it is less recognized that automobile emissions and gasoline in its raw state (as well as other chemicals) can sneak into a home through vents, under doors, and into the attic via the ceiling to produce long-term health effects.

story. A good illustration of just how easily this can affect an individual, was published by Dr. Randolph. It involves a somewhat amusing tale of a nun who couldn't for the life of her stay awake during mass every morning. She had to be repeatedly nudged by her peers. Dr. Randolph suspected that the warm vapors (consisting, by the way, of not only gasoline, but oil, coolant, and lubricating grease) that were emanating from the automobiles parked in the garage directly under the chapel which had been drifting upward through the floor had something to do with it. Sure enough, when everyone left their cars outside, her embarrassing problem ceased.

Also be advised that high-rise apartments and office buildings can invite problems. It is possible for vapors to travel from parking garages or loading docks up through stairwells and right into every area of the structure. Keep in mind, also, that some especially sensitive individuals have been affected just by the raw gasoline fumes evaporating from the carburetor of a cool engine.

Even when the matter of automobile parking is completely under the control of the individual and the utmost caution is exercised, it is not, unfortunately, always enough. If heavy emissions from gasoline vehicle exhaust (or other pollutants for that matter) are up wind from a residence, it can prove to be sad news for the occupants. Never open windows in that direction. Whenever airing out your house or apartment, wait until the wind has shifted. Then, if temperature permits, by all means make it a point to refresh the air in this way, because some quantity of the vapors, as with any persistent pollutant, will have entered and will linger in the confined spaces. Conventional heating and air conditioning systems cannot be heavily relied upon because their filters are virtually useless for screening anything in the form of vapor, and their air-replacement capacity may not prove as effective as that which can be delivered from a set of open windows on a breezy day (when the wind is "right").

If prevailing winds, however, continually push such heavy pollutants in the direction of your home before they are adequately dispersed due to the closeness of a major highway, for instance, there may be reason for concern if you are chemically sensitive. Not only will combustible-laden air enter open windows and sneak in to some extent even if they are closed, but it can be pulled in by heating and air conditioning systems.

Best Locations for Avoiding Exhaust Fumes

Sadly, some may be forced to relocate because of this problem. If this becomes necessary, there are some things to be considered. Of course, it is common logic to avoid any dwelling located adjacent to a highway, a main thoroughfare, or a major intersection. But there should be a good safe distance between a home and such sites. Although local vehicle pollution diminishes sharply with distance, a rule-of-thumb for the highly sensitive is, if you can hear the din of traffic from the premises, you're too close. Large parking lots of churches, schools, shopping centers, or the like in the vicinity are not to be overlooked either. They not only provide a refuge for a large volume of cars (and their exhaust), but often invite idling diesel trucks and busses. Obviously, living near airports is out. The emission of one jet upon takeoff is equivalent to that produced by as many as 7,000 passing cars![6]

For those who have become especially sensitive, even a nice, quiet neighborhood in the city could prove less than acceptable, and the only safe place might seem to be the country. But before making a hasty retreat from the city, bear in mind that even the country might not be a haven for long. There have been sad incidents in which chemically susceptible people have gone to the trouble and expense to plan and build a home on a pleasant, out-of-the-way site only to soon discover with horror that a construction crew is beginning work on a new highway not far from their back door. If considering a move to the country, don't make any firm commitments until you inquire from county authorities if any highway construction (or industrial plant or factory construction, for that matter) is planned for the area you have selected.

When choosing the location of a home, make certain that as many windows as possible face the prevailing winds head on in an area free of chemical emissions. Also choose the highest elevation in the vicinity, not a valley where stale air can become temporarily trapped.

At any rate, remember not to travel unnecessarily or otherwise expose yourself to exhaust fumes if you find yourself sensitive to them. And to dilute any trapped emissions, offer your indoor domain some fresh air whenever it is available. Those who take precautions might very well be surprised at how much better they feel, and may even discover that all the time, they have been "car sick" in the house.

HAIR, DANDER, AND FEATHER WOES

Domestic animals are extremely popular in the United States. Over 100 million are kept as pets. Although they can provide pleasure, ease tension, and promote love, there's no telling how many sneezes, runny noses, itchy eyes, and other symptoms have been brought on by their hair, dander, and feathers. No question about it. These ubiquitous substances are one of the leading causes of allergic diseases, particularly those of a respiratory nature.

Common Furry Domestic Animals and Allergies

Dogs

As far as our furry friends are concerned, dogs make one of the largest hair and dander contributions. The dandruff is especially bad to sensitive individuals because it can easily break down into tiny flakes. In addition, it absorbs sebum, the oily substance that is secreted by the dog's sebaceous glands, which itself has been known to initiate symptoms. Still, many have been affected by the hair. As in the case of other such animals, the protein matter of the hair shaft, largely keratin, is what allergists point to as the instigator of health problems, but there is room for argument because the hair, too, is coated with sebum. And whether Chihuahua or poodle, cocker spaniel or German shepherd, short-haired or long-haired alike, there is no hard evidence that one breed is worse than another, although it is possible for an individual to be affected by just one breed. (Note: Some dogs such as poodles may produce somewhat less of a plight in the sense that they shed less, therefore leaving less hair behind. But in an allergenic sense, their hair is generally no different, than any other dog's.)

Cats

Cats, however, are generally worse offenders. Their hair is finer and softer than a dog's, and it sheds more readily. Moreover, there is an allergen in the saliva of a cat which is deposited on the coat during grooming. Both the hair and dander, which also contain sebum, tend to linger in the air or become attached to home furnishings indefinitely. In addition, these substances are renown for clinging to clothes. Unlike those of dogs, breeds with longer hair have a tendency to be more allergenic than the others. Some severely sensitive people have been known to have an allergy attack, a flare up of asthma, or other such problems even when entering a room or building in which a cat had previously been an occupant. Up to 30% of those with chronic respiratory problems, particularly asthma, have been found to have an allergy to cats.

The tricky thing about this situation is that reactions can start happening to anyone at anytime. A person could have lived with a pet for years before symptoms reveal themselves. It just takes some longer to become sensitized. Also, a pet could have been long removed from the premises and a susceptible person can still suffer because of the lingering residue.

Although adverse human reactions could be caused by a flea collar or flea powder (as discussed in chapter 6, *The Chemical Crisis*), or perhaps medication that has been applied to the coat, one still may face the possibility that reactivity to hair and/or dander will force him or her to give up a pet dog or cat.

Pet Tips

Here are some tips, however, that might save one that grief. At the very least, the pet should be kept off all carpet. By rights, it should be banished to the yard. Not only will this move improve the condition of the inside environment, but it will also curb the animal's shedding. This is because light has a bearing on the shedding process. In outside conditions, a dog or cat will only shed 2 times a year—in fall and late spring. If they are allowed to remain in the house, where electric lights are invariably turned on as soon as it gets dark, the shedding will continue year round.

Additionally, it is a good habit to always wash hands after petting and let the grooming be done by someone in the family who is not sensitive. Your animal should be brushed 2 or 3 times a week. This practice works wonders for keeping them clean. Use a soft brush, though. Stiff ones will scrape the skin and release more dander. And, as might be expected, long-haired animals need more attention in this respect. It is more difficult to remove the dander and excess hair from them.

Another idea is to wipe the pet down with a wet towel. This is a good way to remove much of the saliva and loose hairs. Even a moist hand rubbed over a short-haired animal will pull out a surprising number of hairs.

Also, the pet should be shampooed at least every other week. But use a veterinary shampoo, as the animal's sensitive skin is not designed for the harsher formulas produced for human scalps. This measure will, of course, also place less strain on sensitive airway passages, as stronger shampoos will release more chemical vapors.

The coat can even be made nonstatic by rinsing it with a solution of one teaspoon of fabric softener in a quart of water. But watch out for chemical sensitivity. Fabric softeners contain perfumes and other ingredients that are troublesome for some individuals.

Don't forget to wash bedding frequently. This could be done after the animal has been bathed each time. It will be easier to remember that way, and a clean pet will start out fresh with a clean bed.

Here is something else to consider. Most authorities agree that animal dander problems can be minimized by proper nutrition. Going into detail about canine and feline diet is beyond the scope of this book, but pet owners should make sure their animals are getting properly balanced meals. Check with your veterinarian.

Other Animals

Horses can initiate reactions in people, too. Although it is highly unlikely a horse will be found in the house, stables and barns are another thing. As many as 19% of allergic patients are sensitive to horse allergens. The usual resulting ailment is asthma. Some are so sensitive to this kind of animal, even being exposed to horse manure, as in a garden, can set off a reaction. Such sensitive individuals are also often allergic to mules, donkeys, and zebras. These people should obviously avoid all places where horses (and other such animals) are likely to be.

Beware of other pets as well. Rabbits, gerbils, hamsters, ferrets, and mice can be a potential source of trouble. Also, those working with animals in research labs are vulnerable from excessive exposure to guinea pigs, rats, mice, rabbits, or monkeys. Veterinarians, zoo workers, circus performers, pet shop employees, and the like can be victims as well.

Objects Made From Animal Hair

Living animals aren't the only origin of problems either. Although synthetic fibers have largely taken over the role that animal hairs once played in stuffings, clothing, and such, because they are less expensive to manufacture, there are still some items that have escaped this modification. Oriental rugs, like Chinese and Persian, are made of animal hair, including that of the cat and dog. Cat hair can also be found in fur gloves, caps, and slippers. Horse hair is still employed as a binder in plaster, as a stuffing for orthopedic mattresses, and in some brushes, clothing, chairs, violin bows, hats, gloves, shoe linings, wigs, ropes, and twine.

Rug pads, brushes, and antique furniture may contain hog hair. It still may be found in old mattresses and automobile cushions because it was once used extensively in items of this nature. Rabbit hair can be used for things like felt hats, gloves, stuffed toy animals, robes, lining for jackets and coats, imitation mink stoles, and hammers of piano keys. And here's an irony. Superstitious people may be trying to ward off bad luck with the use of a rabbit's foot while all the time it could be producing illness in them.

Goat hair may be contained in some apparel. It is sometimes mixed with wool and is used for the manufacture of rugs, rug pads, carpets, drapes, doll's hair, wigs, blankets, and also inexpensive hairbrushes and paintbrushes. Some know it as mohair, which is technically hair derived from the Angora goat. Age seems to make goat hair environmentally worse.

Oriental rugs (particularly from India and China), chenille carpets and bedspreads, hairbrushes with imported bristles, felt, and some twines and ropes can be made of cattle hair. Camel hair, of course, is used for camel-hair coats and jackets, as well as some brushes. Expensive old shawls were once made of camel hair too, and some are undoubtedly still used.

Of course, any items of this kind in the home can become a problem. But workers in the fur industry can be victims, often developing a condition known as fur asthma, as can hat makers who may have problems with rabbit hair.

As far as wool is concerned, it is an improbable candidate for the production of respiratory allergy because it undergoes such extensive processing. Such items as heavy sweaters and blankets, however, can cause problems since small particles of the pure substance are easily inhaled. Wool blankets, for those so sensitive, may require a sheet or cotton cover on both sides to confine any matter that would normally be released. Of course, wool clothing will probably have to be abandoned.

It is also possible to be allergic to human hair.[1] Barbers and wigmakers should especially beware. And the quantity involved may be negligible, but also watch for false eyelashes, which, although often made from synthetic material, can be produced from real hair.

No one has confirmed that human dandruff is a source of allergy or other disease. But it's something to think about. In one study, over 90% of a group of asthma patients tested exhibited a positive skin test reaction as compared to a normal group who showed no response.[2] While this leaves much to be desired in the direction of establishing proof, it certainly brings into light the possibility that your own or someone else's dandruff may be bothering you.

Feathers

Of course, our feathered friends pose a potential health problem, too. Items such as pillows, quilts, coats, ski jackets, sleeping bags, upholstery, and mattresses can be stuffed with goose or duck. Also don't forget about Indian headdresses or the feather duster. These articles have been known to cause ailments such as asthma and hay fever.

Interestingly, freshly plucked feathers elicit less reaction than old ones.[3] This suggests that the denigration process of feathers is at fault. Another reason may be because, in time, they absorb some other things from the air, such as mold or bacteria. Also, although an allergy to the feathers of only one species of animal may exist, if an individual is allergic to one kind of feather, he or she is usually allergic to other kinds as well.

One might reduce any problems caused by covering each side of a feather-stuffed pillow, for instance, with a pillowcase. But such a product may have to be given up. Beware, however, of synthetic material as a substitute. Its chemistry can create health problems of another nature. Try to determine whether or not you are sensitive to an item of this kind before you obtain it, or it could prove worse than the item it was substituted for. The safest replacement is usually cotton batting.

Actually, feathers on living birds fraught with dander and attached to the oily portion of the skin are more apt to initiate allergic symptoms than the plucked feathers used for stuffing.[4] Canaries, parrots, parakeets, myna birds, and pigeons are pets that might be an unfortunate source of trouble. Not only can a child be besieged with constant respiratory problems because of a pet parakeet in his bedroom, but a farmer can be affected with a flare up of sinusitis every time he goes into his henhouse, or an elderly gentleman might suffer an asthma attack even after his daily routine of feeding pigeons in the park. One woman's asthma was reported to have been caused by sparrows which had been nesting in vines beneath her bedroom window.

And while on the subject of birds, it is only fair to mention that minute particles from the feces and nasal secretions of birds like those of parakeets and pigeons have been known to trigger allergic responses such as hypersensitivity pneumonitis (involving inflammation of alveolar walls and bronchioles), or to transmit psittacosis (better known as parrot fever). Both can be life threatening.

The only hope for those stricken owners of pet birds is to clean cages often and keep the pet in the least occupied room. This, however, is usually not enough.

No matter what kind of pet is involved, some may be tempted to experiment by leaving it with a neighbor, a friend, or the nearest kennel. But don't make the mistake of thinking this will be a conclusive test. The house must be thoroughly cleaned during this period in order to eliminate all the residual contaminants. This especially applies for dogs and cats because they are bigger and are usually allowed to roam free, where birds, hamsters, and other such animals are generally confined to cages. Although it is not to say that any of these creatures aren't directly or indirectly capable of disfavorably altering indoor conditions, fish, turtles, chameleons, or even a snake (provided it is nonpoisonous, of course) place less of a burden on the environment.

Generally, the safest thing to do if you find that your bodily sensitivities fall into the hair, dander, or feather category, is to avoid as many living furry and feathery animals or items made with animal hair as possible. The more of these things that are removed from your life, the better off you will be.

WHAT ABOUT BACTERIA?

The subject of bacteria and their contribution to illness could fill volumes. Although it is a topic that is more suited for medical or biology textbooks, it deserves a measure of attention within these pages, since it, too, is ever-present in the indoor air.

There is a vast amount of detail about these microorganisms still shrouded in mystery or infectious illness would be a thing of the past. Man has yet to conquer the common cold (actually brought about by a virus, which is another kind of entity, typically smaller, that routinely invades bacteria). Just when he thinks he has labeled every ailment, up pops a surprise like Legionnaire's disease or AIDS (another malady caused by a virus). When individuals are treated for infection, they don't always respond as expected due to a host of unanticipated intervening factors which often play havoc with routine procedures.

At that, man has subdued many dreaded diseases, such as bubonic plague, typhoid fever, and leprosy (Hansen's disease). And with the help of drugs, he does possess reasonable control over infection. Non-infectious chronic ailments, many of which are brought on by indoor pollution, are what is now his major nemesis.

Not every species of bacteria are instrumental to disease anyway. Far too many of us have the mistaken impression that we are surrounded by a multitude of dangerous organisms, which are constantly at the ready to assail the human body, and are staved off only by man's growing chemical "arsenal." In reality, although bacteria *are* plentiful, most are harmless—and many are directly beneficial to man. Some, for instance, are necessary for fermentation. Others are needed to convert atmospheric nitrogen into nitrates which maintain the fertility of the soil, where most bacteria, incidentally, are found. Still others are regular residents of our body, some of which are guilty only of aiding certain human biological processes. And who knows? Perhaps someday, man will discover positive purposes for other kinds of bacteria as well.

How We Succumb to Bacteria

Actually, exposure to any one or more varieties of harmful bacteria does not necessarily mean illness. It is what we do to lower our resistance that enables these microorganisms to have a detrimental effect upon us. Astronomical numbers of bacteria are ejected into the air at every cough or sneeze when a tissue or handkerchief is not used. They land on furniture, floors, shelves, toys, eating utensils, hands, food, and other objects, and many remain airborne (some surviving for considerable periods if conditions are right) in proximity of others within stuffy rooms where some are inhaled. Depending on the kind of bacteria involved, if the body is weak, as from undergoing stress, recovering from surgery, taking in the wrong kind of foods, or becoming chemically sensitized, it may succumb.

Be assured, however, that physical contact is of greater concern because very few serious diseases result from the conveyance of bacteria through the air. Most disease-producing bacteria (and viruses too) generally cannot survive long in the atmosphere. But infectious ailments, especially of the less dangerous variety, do occur because of airborne microorganisms—and far more often than we would like.

Where Bacteria Thrive

The primary reason for this is because, although air is not their favorite medium, the inside atmosphere can be constantly replenished with bacteria from other sources. They (along with mold, which is capable of survival in the midst of even less moisture) can find an adequate habitat in damp areas, as around toilets, within refrigerators, on damp walls, and in poorly maintained air ducts, air conditioners, humidifiers (in which some researchers are of the opinion bacterial contamination could be near 100%), dehumidifiers with drip pans rather than piping to divert moisture outside, cooling water towers, or any other place where moisture is present.

Bacteria can also be found thriving in poorly maintained air purifiers, wooden chopping boards, wooden tables, decaying wood, plants, and sometimes even glue.

Bacteria, like other contaminants, can also be brought into the indoor environment on shoes, clothing (on or off the person), hair, pets, and insects, where they may survive just long enough to infect someone. In addition, they can hitch a ride on dust particles.

It is also possible for bacteria to infiltrate the home in odorous air escaping from the sewer line via unused or seldom used sinks, bathtubs, or toilets. (Water must flow at least occasionally through these devices to restore the water seal in the trap.) The same goes for basement sumps and floor drains if they are connected to a sewer. This problem can likewise arise if plumbing has become damaged from corrosion or house settling.

Unknown to many, the range of temperature within which bacteria can grow and carry on their functional activities is fairly limited, being approximately 80° to 100° Fahrenheit for most species, although these figures can go notably lower or higher for some varieties.[1] (The ones responsible for human infection, appropriately enough, reproduce most efficiently at 98.6°, the normal body temperature.) But they are capable of enduring a much broader span, particularly below 32° Fahrenheit (some even having been found alive after being frozen in ice for years) and, in some cases, up to the temperature of boiling water (the reason an autoclave, which employs pressure and a higher temperature, must be used to guarantee sterility). So if the species is harmful and continues to survive in a certain location, there is always potential for the creation of infection.

But in order to survive in this broader temperature span, they require not only moisture (the composition of an active bacterial cell being 90% water), but other necessities as well.

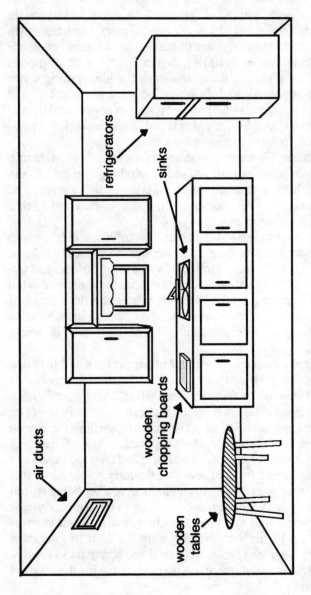

Places in the kitchen where bacteria flourish. Although we usually consider the bathroom the most likely candidate for harboring bacteria, the kitchen can also present its share of problems.

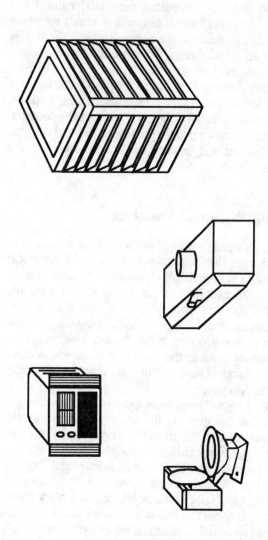

Other areas where bacteria are prone to multiply. Humidifiers, toilets, air conditioners, and cooling water towers are excellent examples of objects which are notorious for supporting bacterial growth.

For one thing, they need food. Some are "fussy eaters," but others are capable of living on a wide variety of nourishment, including wood and other cellulose materials, natural and synthetic resins, and even substances such as man's very own food products, not to mention the living tissue of the bodies they invade. In addition, most species are fond of darkness. Sunlight often retards their growth.

If one person inhales some that are functional or even dormant, and coughs, sneezes, or even talks, laughs, or sings, whether he becomes infected or not, he develops the potential to infect others. The more people present in the vicinity, the more will be inhaled and exhaled, and the more chance a greater number can be infected.

Threatening Bacterial Diseases

Legionnaire's disease (Legionellosis), a pneumonia-like ailment, which was first recognized at the 1976 American Legion convention in Philadelphia's Bellevue Stratford Hotel, proving fatal to 29 people, is a good example of a serious disease brought about by conveyance through the air. The microorganism responsible, now called *Legionella pneumophila*, was discovered growing in the air conditioning system and was later recognized as the culprit in hundreds of other cases, including those in hospitals in Los Angeles, Memphis, and Burlington, Vermont.

Humidifier lung, believed to be induced by the actinomycetes bacteria that can grow in the water of idle humidifier systems, is another serious illness transmitted by the air. Accompanied by fever, chills, a cough, headaches, and indigestion, it swells the tissue of the lung and impairs breathing ability. Numerous office workers have been afflicted with this, especially at the beginning of a work week, when a humidier is turned on following a weekend rest. There have even been reports of incidents in which symptoms of this ailment did not occur until long after workers had returned home.

A rarer example of disease transmitted by the air is psittacosis (parrot fever), known also as ornithosis, a virus which originates from microscopic particles of the feces and nasal secretions of infected birds, particularly parrots, parakeets, pigeons, turkeys, ducks, and geese. It is characterized by headache, insomnia, fever, chills, abdominal pain, vomiting, sometimes delirium, and especially coughing.

Tuberculosis, strep infection, meningitis, diphtheria, scarlet fever, whooping cough, and pneumonia are other prime examples of threatening diseases caused by airborne bacteria.

Conditions That Encourage Bacterial Infections

Office buildings and other public places which house large numbers, especially with low ventilation rates furnish an excellent environment for the transmission of disease in this fashion. Some business offices have become the scene of epidemics of other dangerous or potentially dangerous diseases such as measles and influenza, (both viral infections) and even Q fever (a disease caused by a rickettsia, which is another kind of microorganism comparable to the size of a virus).

A seemingly innocuous office, in fact, can prove to be something much more when it comes to airborne bacteria, let alone chemical emanations and inorganic particulate matter. In one instance, in a large office complex in the southwestern U. S., health investigators discovered to their horror that the air conditioning system, which employed an open-spray water unit, was pulling in bacteria. Analysis of the water turned up 14 different types of harmful organisms.[2] Needless to say, such examples demonstrate the importance of keeping air conditioners, vents, and other such apparatus clean in the home.

Controlling Bacteria

Heat in adequate amounts, of course, destroys bacteria—and so will gamma radiation (which is lethal to humans as well), certain chemicals (not good for the chemically-sensitive), and ultraviolet light. Some might consider ultraviolet light the answer for keeping the inside air sterilized, but there are a number of drawbacks that make this method less than completely effective. First, the bacteria must come in direct contact with the light. Then it must remain long enough to absorb the amount of energy required to complete the job. In addition, dust and other particles will absorb and diffuse much of the energy necessary for its destruction. Even a minute amount of dust coating the light bulb is sufficient enough to diminish its output to the point of ineffectiveness. Also, humidity will impede the process of disinfection. And it might be added that ultraviolet lighting is likely to contribute ozone.

The best methods for reducing the bacteria population in the indoor atmosphere is by employing many of the suggestions outlined previously in these chapters.

Ventilation is an excellent way to control airborne diseases indoors. As in any case, it serves to dilute the harmful matter. Opening windows can be especially beneficial after a rain, at which time all of the outdoor bacteria has been washed out of the air and has settled into the soil.

Cleanliness is a must. Not only can living bacteria be swept back into the air from a settled position, but a contaminated hand which is brought to the nose or mouth after coming in contact with a less than sanitary surface can cause infection. Even flies, roaches, ants, and other insects are capable of becoming a transmitter of diseases such as cholera as a result of walking across such areas.

Remember, borax and water serve as a mild disinfectant. It can be used by the chemically-sensitive for scrubbing toilets, refrigerators, tile, or most other places that require such attention. Clothing and bedding should be washed regularly because the lint and dust arising from them are such an efficient means of disease transmission. Borax can be used for this purpose as well. Also make it a firm habit to wash hands before eating and after using the bathroom or blowing your nose.

Don't forget, by the way, that bacteria can settle on food. To avoid intestinal diseases, food should be thoroughly cooked and promptly consumed or covered and refrigerated. (Moist foods with little acidity, such as puddings, soups, and pies, are the best medium for the growth of microorganisms.) Dishes should always be scalded, and garbage should be removed often.

Air purification is effective. Although any system will catch some microorganisms, the most efficient kind is a HEPA filter, since it is capable of screening not only dust particles which might be carrying potentially-harmful microorganisms, but matter smaller than that of the average-size bacteria.

More efficient vacuuming equipment, particularly in the form of a central vacuuming or water-trap system, will help considerably. The more dust or other particulate matter that is eliminated, the less chance for dissemination of bacteria, viruses, and other contaminants.

Humidity control is also important. Many bacteria ride along the air currents in moisture and especially thrive in a humidity level above 60%. So the less of it there is, the less chance that bacteria will survive. For homes or any other structures which are plagued with excess moisture, a dehumidifier should be put to use.

Be advised, however, that microorganisms will not necessarily die just because the humidity level is low. In fact, the unseen droplets of moisture that surround much of the particulate matter in the indoor atmosphere are even smaller under this condition, which makes them lighter and therefore more likely to remain airborne. That's the reason dust control can be so important.

In taking such measures you will not only reduce your chances of contracting infections such as colds and flu or perhaps a worse illness, but, in many cases, you will rid yourself of other contaminants that can be the cause of chronic ailments.

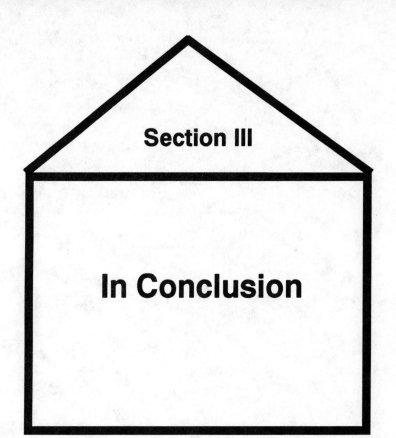

Section III

In Conclusion

OTHER FACTS NOT TO BE IGNORED

Hopefully, you are now familiar with the many contaminants that are found indoors, and have developed a better understanding about how illness can be induced by them as well as how to control or eliminate them. But before closing the book on the subject, here are some additional tidbits that should not go unrecognized.

Carbon Dioxide

Firstly, no attention has been given to carbon dioxide. Always present in the atmosphere, it is not a toxic substance, at least in small amounts, and in fact, we not only take in a limited degree of it every time we inhale, but normally release a content of approximately 4% of it in our exhalations. It is responsible for "that stuffy feeling" that sometimes occurs inside homes and public buildings. Since it is a human and animal by-product, it is rarely considered a pollutant and usually poses no problems, but there are extreme cases which make matters different. When carbon dioxide has increased to more than 5% of the atmosphere, breathing will become labored. This can result in a headache, nausea, shortness of breath, drowsiness, and poor judgment. A 10% ratio is the upper limit that can be tolerated. Higher concentrations will depress respiration and ultimately cause unconsciousness and death. Obviously, if such a problem should become a concern, it can be remedied with any reasonable amount of ventilation.

Gas ranges and heaters, when in use, render carbon dioxide. In fact, any flame will displace some amount of oxygen with carbon dioxide, even that which is produced from a match or candle. This would contribute little to indoor pollution, but in the case of the candle, the vapors produced from its content must be considered. Some candles are made from paraffin, which consists of hydrocarbons obtained from the distillation of petroleum. Also, scented candles contain chemical additives. If frequently engaged in the act of burning candles, the only acceptable kind is that which is made from beeswax.

Leaking Batteries

Another easily overlooked potential chemical problem concerns leaking batteries. It is not uncommon to have one or perhaps more batteries stored on a shelf or in a cabinet that have gone bad in this way. Not only does this highly corrosive substance ruin surrounding materials, but often the odor can linger even after the affected area has been cleaned. There have been no reported cases of symptoms attributable to this, but neither is there any reason to believe that such an item couldn't cause health problems, if it does nothing more than provide a contributing factor to illness. Check around for old forgotten batteries and examine them. Even if they appear intact, avoid future problems by testing them. If they are weak or dead, they should obviously be discarded.

Activities That Generate Particulate Matter

Watch out, hobbyists. Glue, paint, fillers, and other such paraphernalia are bad enough, but when filing, sanding, planing, or drilling, the particulate count of the surrounding air zooms upward. Tiny bits of wood, metal, plastic, or other material can be readily breathed. Wood dust from oak and cedar are especially bad about bringing on asthma, although other kinds can be just as bad.[1] Of course, metal and plastic dust are also very harmful to the lungs. Just as in the case of any particulate matter, a high enough concentration will reduce lung function and increase the load on the heart. Confine such activities to some appropriate place outside the house—and when working, always wear a filter mask along with eye safety glasses and gloves.

Even erasers release some matter. It may be a minor point, but it certainly won't hurt to take the time to dispose of the debris properly, not distribute it in the air by blowing it haphazardly off the paper, particularly if the eraser is in frequent use. Chalk dust, a variety of limestone or calcium carbonate, can also fill the air with particles the lungs don't need.

Fiberglass Filters

Conventional fiberglass furnace filters, just as any other products containing such a material, are capable of releasing glass fibers into the air after they have aged. In addition, most air conditioning and heating filters, fiberglass or otherwise, are treated with oils, adhesives, plastic resins, and/or hexa-chlorophene.[2] To avoid the problem of chemical and particulate dispersal because of these factors, untreated metal mesh filters should be used. (Simply spraying them with olive oil will increase their ability to trap dust particles.) Of course, as previously explained, HEPA and HEPA-type filters are an even more efficient choice in safely extracting particulate matter from the air.

Freon Leaks

Air conditioners, refrigerators, and heat pumps can be an unexpected source of trouble. It is an established fact that the leakage of freon around pipe joints can result in serious illness. If you are unable to locate the source of your symptoms, it could lie here. Make certain these apparatus are functioning to capacity, and arrange for repairs if necessary. (Note: Don't forget that automobile air conditioners are capable of leaking freon too, and can not only contaminate the vehilce, but closed garages as well.)

Emissions Produced by Electric Motors

Motor-driven appliances such as electric can openers, food processors, sewing machines, and drills release some degree of ozone and oil fumes.[3] This has lead to a number of cases of ill health. It is suggested that for the highly sensitive, such items be used sparingly and with sufficient ventilation, or used on a terrace or patio. Also note that fans, refrigerators, air conditioners, and the like fit this category as well. Some devices have permanently sealed motors and these are certainly more desirable.

Fluorescent Lights

Old fluorescent "rapid start" type light fixtures, generally those manufactured before 1978, have capacitors in their ballasts (electric starters) which contain PCBs and asphalt products that can leak. To check, remove the cover from the light and look for a small metal box. It is usually mounted between the tubes. If there is any sign of black or oily residue, discard. These lights have been known to raise PCB levels thousands of times the outdoor level. PCBs are an eye, ear, nose, and throat irritant and over time are capable of causing severe liver damage. They are a suspected carcinogen and can build up in the body to cause fetal death and mutation. Fluorescent lighting, as well as mercury enhanced light bulbs, also give off ozone.

Other Materials That Emit PCBs

Electrical transformers, water-proof adhesives, and even NCR paper also may contain PCBs. There is no evidence that the amount found in NCR paper is enough to be a hazard, but the former two can be a different story. This holds especially true with transformers because, like the capacitors in fluorescent ballasts, they may become damaged and begin to leak this substance.

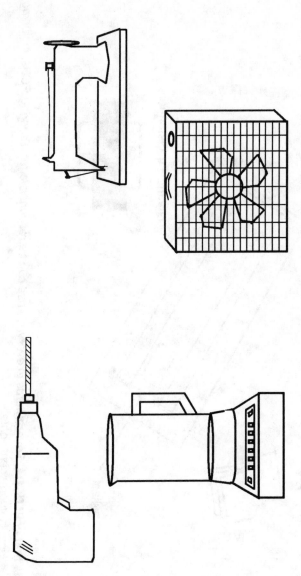

Devices that emit ozone and oil fumes. Any devices which employ electric motors produce ozone and oil fumes.

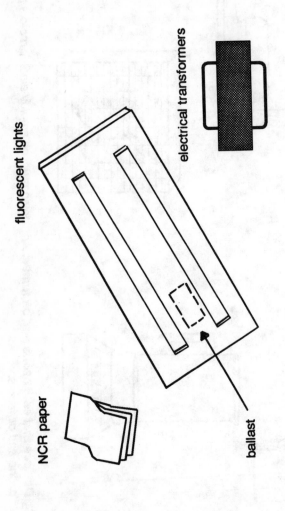

fluorescent lights

NCR paper

electrical transformers

ballast

Items that emit PCBs. *Several items in everyday use can be responsible for dispering PCBs, including the ballasts in fluorescent lights.*

Other Items That Can Lead to Allergy

There are other materials that can produce respiratory symptoms. Kapok is a plant substance that is ground up for filler in pillows, sleeping bags, and doll and toy animal stuffing. Jute is sometimes found in rug and carpet pads and coarse ropes and twines are made from jute. Hemp is used for ropes, twines, and cables. It is even an ingredient in some plastics. Sisal is a Mexican plant similar to hemp used for making items such as rope, sacking, and twine. Cotton linters (cotton in which some of the seeds were not removed in the ginning process) are used in the manufacture of mattresses, cushions, pads, upholstered furniture, book bindings, sleeping bags, mustard plasters, and linoleum.

Burlap is often made from hemp or jute. It, too, has been known to cause respiratory problems. Watch out for burlap bags. Also note that burlap is used to cover the webbing or springs of furniture to provide a smooth surface on which the stuffing can be easily spread. Burlap is also used under the fabric to help protect it from pokes. This material is also used for wall-hangings and curtains.

Some telephones have a wad of bacteria-killing cotton in the receiver. Unscrew the mouthpiece and check. If you spend a lot of time on the phone (at home or at the office), it would be a good idea to remove it. The chemical used to destroy the microorganisms could be playing a part in destroying you, too. A number of chemically-sensitive individuals have been adversely affected just by this one small hidden object.

Another Word About Automobile Air Conditioners

Automobile air conditioners will throw out an abundance of dust when they are first turned on in the summer. They can also be a source of mold. It is a good idea to have these devices and the vents cleaned every season before using them. Also, the condensation water should be removed often to keep the system free of mold. The vents should also be checked before firing up the heater in the fall.

More on Pets

Sometimes, not only a pet's saliva, but its urine can cause adverse reactions in humans.[4] Pets may not always be the source of a person's poor health, either, but they can be affected by pollution just as humans can. Flea collars, flea powder, or plastic bowls could prove to be a possible problem, as well as any other pollutant. Some pets have even been known to die because of continual exposure to substances such as formaldehyde. This suggests, of course, that any pet who spends most of its time indoors and becomes chronically ill may be heralding trouble for its master(s).

Fingernails and Toenails

Human hair and dandruff have already been noted as a possible source of symptoms, but one might also consider fingernail and toenail filings and clippings. Although there is no proof of such a thing causing illness, human nails have a similar composition to that of hair and should not be ruled out as a culprit.

Converted Rooms

And here's something else to consider. When transforming a garage into an extra room, remember that the garage has most likely played host to a storehouse of chemicals, especially gasoline, and there have probably been spills from time to time. Oil and grease leaks from cars have no doubt soiled the concrete floor. Some of these remnants may continue to outgas and will be worse when the structure is permanently closed in.

Airing Out the House

Airing out a house is a good way to dilute indoor pollution, but it is important to remember it should not be done in certain situations. Always note the direction of the wind and watch for nearby activities that could be sending a high concentration of contaminants your way. Some examples are barbecuing (accompanied by lighter fluid vapor as well as products of combustion), trash burning, yard mowing (which not only contributes chemical emissions, but stirs up grass and weeds), road or parking lot resurfacing, pesticide spraying, or painting. Of course, one should never welcome outside air indoors during times of heavy pollution, but when such episodes have ended, the house should definitely receive a dose of fresh air to clear out any contaminants that have made their way inside.

More on Plants

It has been mentioned previously that spider plants serve to filter out some of the formaldehyde in the air. But golden pothos, philodendrons, and snake plants will help in this regard too. In addition, one can combat benzene to some extent by keeping chrysanthemums and gerbera daisies. For air purification in general, try aloe and banana trees, Chinese evergreen, dracaena, English ivy, mother-in-law's tongue, peace lily, and reed palm.

Landfills

Don't ever build a home on soggy soils or landfills. Perpetually moist soils, of course, invite mold. Landfills are likely to include trash, dander, carcass particles, and other harmful matter. This attracts insects and rodents, and also encourages mold growth. Moreover, decaying landfill will give rise to gas pollution. There have been hundreds of victims who have built on industrial landfills and have subsequently experienced symptoms ranging from chronic respiratory illness to miscarriages and birth defects as a result. Who does not recall the problems presented at Love Canal?

Pollutants in the Workplace

Employment in a chemical factory or a coal mine, of course, entails obvious hazards, but there are a number of other occupations which present their prospective pollution problems. Here are a few examples.

Shipbuilders, brake mechanics, and construction and cement industry workers have the potential to succumb to asbestos. Barbers can be affected not only by hair, but by chemicals they use on the hair from products such as shampoos, tonics, and the like. Wheat and rye flour has been known to plague bakers. Sawdust can certainly be a factor for carpenters and cabinetmakers, as can horse hair and dander for a racetrack employee. One may become sensitized to bleaching agents such as oxalic acid that are employed in the dyeing and bleaching of apparel. Printers constantly go unprotected from the strong content of the inks that permeate the air. A furrier does not only have to be concerned about the possibility of developing a sensitivity to fur, but also to the chemicals used in the industry, including benzene, copper sulfate, lead chloride, potassium dichromate, arsenic, and mercury.

Even doctors and nurses are exposed to nitrous oxide, chloroform, ether, and other gases which are used for anesthesia. Lab technicians, too, are exposed to an array of different chemicals. Agricultural workers can be affected by pesticides and the additives in fertilizer. Chalk dust can become a foe of teachers. Even fine powder from items such as the castor bean can fill the air of a mill where castor oil is being manufactured and cause asthma or other respiratory conditions. Coffee grinding and processing factories, the site of coffee bean dust, may become a problem for some. Silica particles are often an enemy to pottery or tile makers. Keep in mind, too, that most of these contaminants can be unwittingly brought home daily by the bread winner to affect an infant or the rest of the family.

Of course, it has already been mentioned that offices can be a potential threat to health. This problem is often compounded because many building windows are not designed to open at all, thereby severely limiting the opportunity to ventilate. In addition, just like the garage that has been turned into an extra room, offices that have been converted from onetime factories or abandoned warehouses might have absorbed strong chemicals. Such sites could also possess ventilation systems that under perform because they were never intented to accommodate a large body of people for extended time lengths. It might be a difficult decision, but if the situation applies, sacrificing a job is better than ruining health.

Some Additional Considerations

And here are a few more words of warning. (1) Those who have had furniture in storage, beware! A common procedure for storage companies is to fumigate it, hence another possible problem for the chemically susceptible. (2) Although rarely used these days, steam heat can release chlorine vapors into the air from treated water. (3) Another source of carbon monoxide that usually goes unnoticed is the exhaust of ice cleaning equipment in skating rinks. It can produce high concentrations of this gas. (4) And believe it or not, old ironing board covers and hair dryers can contain asbestos.[5] In addition, asbestos has been found as a contaminant in inexpensive cosmetic products such as lipstick and powder.

Light, noise, and even electromagnetic radiation can also be considered indoor pollutants. Fluorescent lighting, for instance, emits ultraviolet light and operates with a somewhat more of a flicker than most other kinds of lighting. Although most experts say that the white plastic cover substantially reduces UV light and the glare brought on by the flicker, some have suffered eye strain, fatigue, nausea, irritability, and headaches that may have been produced by one or both of these effects. Eye strain may be possible with any type or arrangement of lighting since no artificial illumination can duplicate sunlight perfectly. The rule-of-thumb, no matter what type of lighting is used, is to keep it above eye level and in front of one's working position.

Playing stereos and TVs loudly, or the frequent use of heavy motor-driven equipment such as power tools, or even home appliances, can contribute to tension, be a psychological menace, affect the heart and blood pressure, and cause ear damage. Excess noise also burdens unborn babies and reduces the attention span and learning ability of children. It is always good policy to keep TVs and the like at a reasonable volume level and not use any items which make excessively loud noise any more than necessary. It may seem drastic, too, but wearing ear protection never hurt anyone.

And there are those who believe that the low-level exposures that are produced by electromagnetic transmissions from radio and TV stations, the wiring and electronic circuitry of household appliances, or even the house wiring itself can be a source of ill health, although there is little to support this theory at the present time. Since *high* levels of electromagnetic radiation have been known to cause neurological damage, cataracts, and other ailments, for good measure it would be best to avoid heavy usage of items such as computers, television sets, and microwave ovens. This especially holds true for children and pregnant mothers. Also, some experts recommend that no one live closer than 500 feet from high voltage power lines.

One should now be fully acquainted with most every kind of pollutant that often proves hazardous to a person's health in the indoor environment. For a general, but quick reference to the major offenders, their sources, and the symptoms they are likely the elicit, see Table 1 in Appendix A.

CONSIDERING THE FUTURE

The foregoing information is not by any means complete. No publication on a subject of this nature ever really is. More disturbing facts regarding indoor air pollution will undoubtedly surface from time to time, prompted largely by what seems to be, sad to say, a down trend in the general health of the average American.

As public awareness increases, though, it will shine an encouraging light on the situation, because where there is acknowledgement of problems, there is inspiration for solutions—solutions which inevitably tag along behind the problems.

Some recent developments by public, industrial, and governmental agencies indicate that a trend toward improvement is well under way for indoor, as well as outdoor air, at home and abroad.

For instance, the EPA, in a report that has been required by congress thanks to the Superfund Amendments and Reauthorization Act of 1986, has called for extended research on the health consequences of chemical contaminants found indoors and ways in which to reduce exposure.

The EPA has also set up training programs for state and local officials as well as contractors interested in the diagnosis and reduction of indoor radon problems. In addition, it has prepared a pair of informative booklets: *A Citizen's Guide to Radon: What it is and What to do About It* and *Radon Reduction Methods: A Homeowner's Guide*. Meanwhile, congress has enacted the Indoor Radon Abatement Act of 1988, which requires all federal agencies that own buildings to test for radon.

Florida has a law of its own which lists a number of stiff regulations, directed to home builders, for keeping radon out of new houses. In other such matters, the Bonneville Power Administration is offering its customers in Idaho, Montana, Oregon, and Washington a substantial discount for the installation of air-to-air heat exchangers to reduce the concentration of radon in homes that exceed 4 pCi/L. In Pennsylvania, a low-interest loan program is available for homeowners whose houses need attention because of radon. That state has also opened a hotline to answer questions. In Sweden, the law demands that a house cannot be built until the lot is tested for radon. Further, one must take corrective measures within 2 years if his or her home contains 14.5 pCi/L or more. The building codes there won't allow new homes from harboring radon levels in excess of 4 pCi/L, while renovated houses cannot contain more than 10 pCi/l.

Manufacturers have shifted their strategy. They are beginning to use less toxic additives in plastics. Recently, in fact, researchers at the Argonne National Laboratory have displayed the possibility of even greater strides by successfully including the starch of potato scraps as a raw ingredient in its production rather than employing the hydrocarbons that had previously been required.[1] Building materials manufacturers, too, are now turning out a line of products with reduced amounts of formaldehyde.

Some states have passed legislation regulating formaldehyde levels in residences. A few other states have banned the sale of unvented kerosene heaters. And Oregon has passed a law which declares that all new wood stoves sold in that state must achieve a reduction in particulate emission of at least 75% from previously sold ones.

Many cities and states have imposed ordinances which obligate an employer to fully inform any employee of the chemical nature and level of the materials in the workplace, in addition to their potential harmful effects.

The state of California has ordered manufacturers of spray deodorants and antiperspirants to decrease the quantity of propellants in their products at least 20% as of January 1, 1991 and achieve an additional 60% reduction no later than January 1, 1995.

Due to pressure by the nonsmoking community, smoking has been prohibited in many indoor public places. Now one may not only find a smokeless environment on public transportation or in libraries, but in such places as elevators, retail stores, schools, hospitals, concert halls, and many commercial buildings. Even the lodging industry now provides nonsmoking rooms. In addition, some magazine publishers are declining the opportunity to sell tobacco-product ads, and congress has proposed bills that would require cigarette companies to include additional warnings on their labels that alert consumers to the danger of addiction, and death from cancer and heart disease.

The British Medical Association is pushing to ban advertisements of tobacco products, except at the point of sale, while the Norwegian Medical Association has urged their government to set a goal for a tobacco-free society by the year 2000. The Medical Association of South Africa is placing much of its effort in not only pushing for a complete ban on all tobacco advertising, but also for removal of all cigarette vending machines and tobacco products on open display shelves in supermarkets. Even in the Philippines, where tobacco is a major agricultural product, there are rigidly observed restrictions against smoking in many public places. In fact, even many smokers concur that public smoking should be curtailed.

The United States has placed restrictions on the very harmful pesticides chlordane and heptachlor. Meanwhile, Japan, Turkey, and the Scandinavian countries have gone a step farther by banning the sale and use of them.

The state of Connecticut has begun requiring pesticide contractors to provide customers with information on the product label of the pesticides they plan to dispense, while a number of schools across the nation have suspended or severely curtailed the use of pesticides.

A May, 1986 law went into effect in New Jersey, which limits the engines of idling vehicles from running for more than 3 minutes. In the meantime, Germany has banned the use of lead in gasoline, and there is a law on the books that asserts the United States must do likewise by the end of 1992. Several producers of photovoltaic cells, moving in another direction, continue to promote solar-powered vehicles.

Due to increased concerns over the pollutants that become trapped inside vehicles, some automobile manufacterers have begun installing air filters in dashboard ventilation systems.

A Canadian government committee has published a report which suggests 30 recommendations for further government involvement in the study of environmental susceptibilities.

Some European countries no longer permit formaldehyde in cosmetics. The Netherlands has adopted standards limiting indoor formaldehyde levels to 120 microns per cubic meter.

The Federal government is considering classifying paints in the same category as prescription drugs.

There now exists a national nonprofit organization called the Human Ecology Action League (HEAL), which was organized at the request of hundreds of patients who had learned their ailments were created by the environment and wanted to share this information with others. HEAL members include general practitioners, allergists, internists, surgeons, psychiatrists, and the like, whose chief purpose is to educate not only themselves, but the medical community and the general public by publishing information about environmental illness.

Another organization is the Society for Clinical Ecology, composed of physicians, scientists, and other professionals devoted to the study of chemical sensitivity, as well as food allergy.

Some hospitals have set up entire wings (Environmental Control Units) for chemically-sensitive patients, in which not only smoking is prohibited, but all chemicals, whether perfume, cologne, hair cream, or those for purposes of cleaning.

Metal termite shields are now required by many codes and city ordinances. This eliminates the threat of one kind of pest, which, of course, reduces the chance that pesticides will have to be put to use later on.

As of February, 1984, an ordinance in the town of Wauconda, Ill. requires lawn care companies to post warning signs during and after pesticide application.

Individuals have also gotten into the act. Groups have been formed such as the Wimberly (Texas) Citizens for Alternatives to Pesticides, inspired at least in part by an incident involving kids on a school playground who contracted hepatitis after they were exposed to chemicals from a nearby crop dusting session. And a number of victims of environmental illness have established isolated communities.

Meanwhile, an inventor in Denton, Texas has developed an oil additive called "Liquid Ring," which is actually capable of internally coating worn engine rings in motor vehicles. The result: a cooler operating temperature, substantially decreased engine wear, improved gasoline mileage, and most importantly, *reduced emissions*. (See Appendix G.)

No one, of course, can predict exactly what is yet to come. But in the end man will improve his environment because he must, to restore and sustain the quality of life he is intended to possess. You can make significant strides by instituting as many of the suggestions outlined in these pages as possible. If enough care and effort is taken, you might very well notice at least some change in your health status, often even if you thought there was nothing wrong with you at all. Once you do, you will never again fail to see the importance of maintaining a healthy indoor environment.

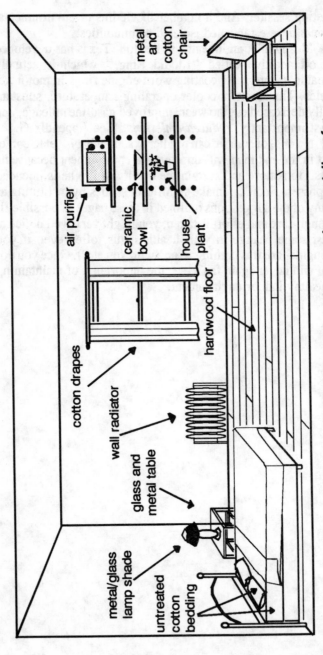

Examples of items which help make a home environmentally safer.

APPENDIXES

APPENDIX A

IDENTIFYING THE CULPRITS AND THEIR PROBABLE HEALTH EFFECTS

The following table offers a listing of the major indoor air pollutants, their primary sources, and the symptoms likely to occur from excessive exposure. Keep in mind that although it supplies a good overview, the information in the symptoms column does not always fit together with the contaminant like precisely cut puzzle pieces. It is technically accurate, but the number of symptoms experienced, their exact nature, and their severity are prone to vary according to the degree of exposure, as well as the duration of that exposure. Other factors include the genetic make-up of the individual and to what extent one gives way to negative emotions. The matter can be further complicated—and quite often is—by frequent exposure to more than one pollutant and/or poor eating habits.

TABLE 1: MAJOR INDOOR CONTAMINANTS—SOURCES AND SYMPTOMS

CONTAMINANT	PRIMARY SOURCES	PROBABLE SYMPTOMS FROM EXCESSIVE EXPOSURE
Alcohols	Cleaners, disinfectants, flavorings, inks, liquor, plastics, rubber, shellac, toiletries	Confusion, dizziness, fatigue, muscle weakness
Ammonia	Blueprint machines, cleaners (especially window cleaners), glue, paint	Headaches; irritation of eyes, nose, and throat; nausea; respiratory distress; salivation; vomiting
Asbestos	Insulation, old plaster, some floor and ceiling tile, street dust, wallboard	Coughing, shortness of breath, tight chest, sputum (can lead to asbestosis and cancer)
Bacteria, viruses, etc.	Air conditioners, air ducts, humidifiers, refrigerators, toilets, unsanitary places	Various symptoms depending on micro-organism involved, most common of which are aches, fever, malaise, clogged sinuses

Carbon dioxide	Chimneys, heaters, and anything else in process of combustion; human exhalations	Drowsiness, headaches, nausea, poor judgment, shortness of breath
Carbon monoxide	Automobile exhaust, chimneys, heaters, tobacco smoke, and anything else in process of combustion; oxidation of household chemicals	Confusion, dizziness, fatigue, headaches, impaired coordination, irregular head beats, loss of appetite, nausea, personality changes, seizures, tightness across forehead, visual disturbances, vomiting, weakness
Chlorine	Bleach, pesticides, plastics, rubber, scouring powder, water	Irritation of eyes, nose, throat, respiratory distress
Dander, hair, feathers	Dogs, cats, birds, or other living animals; objects made from animal fur, or stuffing	Respiratory distress
Dust	Automobile exhaust, building materials, carpets, cooking, fabrics, fires, pollen, pets, soil, tobacco smoke	Respiratory distress

Formaldehyde	Automobile exhaust, building materials, carpets, carpet padding, cosmetics, fabric reatments, gas heat, incinerators, insulation, mouthwash, paint, paper treatments, pesticides, polishes, plastics, tobacco smoke, toothpaste, wallpaper, waxes, wood treatments, and many more	Coughing; depression; diarrhea; dizziness; headaches; heart irregularities; insomnia; irritation of eye, nose, throat, and lung; lethargy; mental problems; muscle and joint pain; nausea; numbness of taste and smell; thightness in chest
Freon	Air conditioners, refrigerators, old aerosol products	Irritation of eye, nose, throat, and lung
Gasoline	Automobiles, lawn mowers, and other such powered equipment	Confusion, depression, dizziness, fatigue, feelings of suffocation, flushing, headaches, irregular heart beats, nausea, slurred speech, staggering
Lead	Automobile exhaust, old paint, tobacco smoke, water pipes	Anxiety, depression, fatigue, headaches, insomnia, muscle pain and weakness, mental problems (degenerates nerves and brain tissue)

Mercury	Paints, contaminated water	Hair loss, headaches, loss of muscle control, nausea, timidity, trembling
Mold	Bathrooms; basements; bedding; carpets; clothing; damp ceilings, floors, and walls; humidifiers; kitchen	Depression, fatigue, headaches, heart irregularities, insomnia, respiratory distress
Nitrogen oxides	Automobile exhaust, chimneys, heaters, stoves, tobacco smoke, and anything else in process of combustion	Burning of eyes, nose, throat; coughing; light-headedness (prolonged exposure leads to impaired lung function and reduced immunity to disease)
Oils	Automobiles, fishing gear, guns, lawn mowers, any operating electric motors	Headaches, dizziness
Ozone	Electronic air cleaners and photocopiers	Blurred vision; chest pain; coughing; headaches; irritation of eyes, nose, sinus, throat, and lung; loss of concentration; shortness of breath; can lead to asthma

Phenols	Air fresheners, cleaners, disinfectants, glue, plastic, waxes and polishes, wood preservatives	Nausea, irregular heart beats, respiratory distress
Pollen	Grasses, trees, and weeds	Sneezing, stuffy nose, general sinus problems
Radon	Building materials, rock, soil, well water	No symptoms (long-term exposure can lead to cancer, fetal and sperm damage)
Sulfur dioxide	Fuel oil and coal burning	Burning sensation in throat and nose, choking, coughing, increase of minor respiratory ailments

Source: Adapted from Stellman, Jeanne, Mary Sue Henifin, Office Work Can be Dangerous to Your Health. (New York: Pantheon Books, 1983)

APPENDIX **B**

COMPOSITION OF CIGARETTE SMOKE

Table 2 presents a good idea of what harmful chemicals are contained in cigarette smoke. Some are products of combustion, while others are directly transferred from the tobacco itself. Carcinogens and toxins are designated, but keep in mind that a number of others are suspected carcinogens, co-carcinogens, mutagens, and/or teratogens. At the very least, most are eye, nose, and throat irritants.

TABLE: 2: SOME TYPICAL CONSTITUENTS OF CIGARETTE SMOKE

Acetaldehyde*
Acetone
Acrolein
Aluminum
Ammonia*
Anthanthrene
Benzene†
Benzo-a-pyrene†
Butane
Cadmium*
Carbon dioxide
Carbon disulfide
Carbon monoxide*

Ethane
Fluoranthene
Formaldehyde*
Hydrocyanic acid
Hydrogen cyanide*
Hydrogen sulfide
Lead*
Methane
Methanol
Methyl nitrate
Methyl chloride
Nicotine*
Nitric oxide

Nitrogen dioxide
Nitrosamines†
Perylene
Phenols
Polonium-210
Propane
Pyrene
Pyridine
Sulfur
Tar†
Toluene
Tung oil*

* *Toxin*
† *Carcinogen*

Source: Adapted from Greenfield, Ellen J. House Dangerous. *(Mt. Vernon, New York: Consumers Union of the United States, Inc., 1987.*

APPENDIX C

THE MANY CHEMICAL PRODUCTS IN USE TODAY (AND A FEW WORTHWHILE SUBSTITUTES)

Table 3 includes a broad category of chemical or chemically-treated products, most of which are fully taken for granted. It isn't to say the list is complete, nor which, if any, could be the culprit in every particular case of illness, but to give light as to how much of a chemically-dependent society we have become and alert you to what items might be at the heart of your health problems. Many of the items are usually associated with skin conditions, but are by no means exempt from blame in regard to symptoms brought on by the inhalation of their vapors or odors. Other products, you might never have even thought of as being air pollutants. For a sound indoor environment, it is recommended that as few of these items as practical be put to use. The fewer there are, the less contaminated the air will be.

For necessary applications such as cleaning, deodorizing, etc., just remember the 3 B's: baking soda, boric acid, and borax. As Table 4 shows, they can be substituted in many general instances. In numerous other situations, just bear in mind that there is no replacement for common sense. A plunger is usually an efficient drain cleaner. (Also sometimes very hot soapy water is adequate.) The old-fashioned tool known as the fly swatter can be used for occasional stray insects rather than running in a panic for the insecticide spray. Traps are effective for disposing of rats and mice. Soap without additives and a little water work wonders for many general cleaning applications. If not prevention, a mild cleaning powder and a generous amount of elbow grease can play its part in taking care of dirty ovens. And there is no reason not to stray from aerosols when there are non-aerosol equivalents for many products. For more detail, refer to Chapter 6, *The Chemical Crisis*.

TABLE 3: POTENTIALLY HARMFUL PRODUCTS FREQUENTLY FOUND AROUND THE HOME

After-shave lotion

Air fresheners

Anti-freeze

Antiseptics

Art supplies

Automobile polish

Baby oil

Baby wipes

Bleach

Brake fluid

Breath spray

Bubble bath

Carburetor cleaner

Caulking compounds

Chap stick

Cleaners (for bathroom,
kitchen, carpet,
windows, etc.)

Cleaning pads

Clothing (chemically-treated,
freshly dry-cleaned)

Cold cream

Cologne

Correction fluid for typewriter

Crayons

Cuticle removers and softeners

Deglossers

Degreasers

Deodorants and
antiperspirants

Deodorizers

Diaper rash ointment

Disinfectants

Drain cleaners

Dyes

Eye Shadow

Fabric softeners

Flea collars

Flea powder

Floor wax

Freon

Furniture polish

Gasoline

Glue

Hair conditioners

Hair creams

Hair lighteners

Hair Remoisturizers

Hair removers

Hair rinses

Hair Sprays

Hair straighteners

Hair tonics

Hair lotion

Home permanents

Incense

Insect repellents

Kerosene

Kitchen aerosols (whipped
toppings, cheeses, etc.)

Lighter fluids (both for
cigarette
lighter and charcoal)

Liniment

Lipstick

Mascara

Metal cleaners (for copper,
brass, silver)

Mimeographed or carbon paper

Modeling clay

Moth balls

Mouthwash

Nail hardeners

Nail polish

Nail polish remover

Newsprint (fresh)

Oils (light machine, motor)

Oil and gas treatments
(for motor vehicles)

Oven cleaners

Paints

Paint removers

Paint thinners

Pan sprays for non-sticking

Paper products (facial tissue,
toilet paper, grocery
bags, etc.)

Patch kits (for inflatables)

Pens (especially felt-tip)

Perfumes

Pesticides (including
hanging insect strips)

Plant food

Pool chemicals

Powders (talcum, etc.)

Printing or photographic
supplies

Protectants (for rubber,
wood, plastic, etc.)

Putty

Record cleaners

Rouge

Rust removers

Sachets (bags holding
fragrance powder)

Sandpaper (glue used
in processing)

Sealants (silicone, insulating
foam, transmission)

Shampoos

Shaving cream

Shoe polish

Soaps, treated (liquid, bar,
powder, clothes, and
dishwashing detergents)

Solder (burning)

Soldering flux

Spot removers (clothes, carpet,
upholstery, etc.)

Spray starch

Tanning lotions

Tape recording head cleaner

Teflon pans

Toilet bowl cleaners (both
bottled and kind that
is placed in tank)

Toothpaste (including
denture cleaner)

Transmission fluid

Typewriter ribbons

Varnishes and stains

Varnish removers

Vasoline petroleum jelly

Wall repair compounds
(plaster, paste, etc.)

Windshield washers

Wood fillers

TABLE 4: SIMPLE SUBSTITUTES FOR SOME COMMON CHEMICAL APPLICATIONS

SUBSTITUTES	BAKING SODA	BORIC ACID	BORAX
APPLICATIONS:	Air freshener Antiperspirant/ Deodorant Cleaner Mouthwash Insecticide* Shampoo Toothpaste	Antiseptic Fungicide Insecticide* Mouthwash	Bleach Cleaner Detergent Disinfectant Fungicide Insecticide* Water softener

* *Effective only for some insects. See Chapter 6,* The Chemical Crisis, *for details*

APPENDIX D

RATING CONSTRUCTION MATERIALS

Below is a list of some of the most common materials used in building construction, classified by the way they rate environmentally. It serves only as a general guidepost since there can be variances in some materials due to factors such as which manufacturer is involved, what production lot is used, and the age of the material. Keep in mind, too, that manufacturers are continually striving to improve their products and favorable changes will hopefully be forthcoming in years to come.

TABLE 5: ENVIRONMENTAL RATING OF COMMON CONSTRUCTION MATERIALS

POOR	*FAIR*	*BEST*
Asbestos	Gypsum board	Brick
Asphalt	(chemical	Ceramic
(shingles,	content varies)	Clay
sheathing paper)	Old linoleum	Formica
Carpeting	Perlite insulation	Glass
Cellulose	Plaster	Marble
Chip board	(may contain	Metal
Fiberglass	additives)	(copper, brass,
(unless used as	Softwood lumber	iron, steel)
insulation)	(may contain	Porcelain
Laminated lumber	preservatives)	Slate
Particle board	Stucco	Stone
Plastic	(may contain	Terrazzo
(brick, resin	additives)	Untreated building
foams, etc.)	Vermiculite	paper
Plywood	insulation	Untreated concrete
Polystyrene insulation		Untreated hardwoods
(Styrofoam)		(maple, oak,
Polyurethan insulation		birch, walnut,
Rubber		ash, hickory,
Tar		cherry)
Treated wood		
Urea formaldehyde		
foam insulation		
Vinyl		
Wallpaper		

Source: Adapted from Rousseau, David, W. J. Rea, Jean Enwright. Your Home, Your Health, and Well-Being. *(Vancouver, B. C.: Hartley and Marks, Ltd., 1988.)*

APPENDIX E

PLASTICS, PLASTICS EVERYWHERE

The following list, by no means complete, is designed to demonstrate just how widely plastics are employed in commercial products and alert you as to the type of items that best be avoided if you are susceptible. It is not intended to suggest that all of these articles will automatically pose a health problem. Some of the items mentioned are made only partially of plastic; others are not always composed of plastic at all; still others you may not have formerly recognized as plastic. In any case, it should offer a fairly comprehensive picture.

TABLE 6: TYPICAL ITEMS COMPOSED OF PLASTIC

Artificial finger nails

Automobile bodies

Bags (for grocery, clothes, etc.)

Band-aids

Book covers and folders

Brooms

Brushes and combs

Building panels

Carpet pads

Carrying cases (brief, luggage, etc.)

Clothing (including hosiery, girdles)

Coatings (dish racks, Teflon pans)

Contact lenses

Containers (kitchenware and those containing commercial products)

Costume jewelry

Dashboards

Dishes

Draperies and window shades

Dust covers

Eating utensils

Electrical boxes and cover plates

Eyeglass frames

False eyelashes

False teeth (and caps)

Fans

Film

Fishing equipment

Flashlights

Flotation devices (as in toilets)

Flower pots

Furniture

Garbage cans and waste baskets

Gears (for clocks, mixers, etc.)

Handbags, purses, and bill folds

Handles (pots, pans, brief cases, etc.)

Heating pads

Housings (appliances, TV's, etc.)

Ice trays

Insulation (wall and electrical)

Lamps

Laundry baskets

Light fixture covers and sockets

Model kits

Mops

Novelties

Packaging

Padding and cushions

Phonograph records

Photographic developing
   equipment
Picture frames
Pillow and mattress cases
Place mats
PVC pipe
Recording tape
Refrigerator parts
Shelving and shelf paper
Shoes
Shower curtains
Shower heads
Skylights
Steering wheels
Sunglass lenses
Tablecloths
Tape (cellophane,
   electrical, etc.)

Telephones
Tile
Toilet seats
Toilet paper holders
Tool boxes
Toys (especially inflatables)
Traverse rods
Typewriters
Umbrellas
Upholstery
Venetian blinds
Washing machine parts
Watch crystals
Waterbeds
Window screens
Writing pens

APPENDIX F

KNOWING YOUR POLLEN

Although normally associated with the out-of-doors, pollen is as much a part of the indoor environment as any other kind of pollutant. Many already know what variety or varieties trigger symptoms in them, often even when they remain inside. Others might not be aware that it is pollen which is affecting them. At any rate, although neither can be considered complete, the following tables are designed to assist you in identifying some of the types of pollen that can elicit symptoms and in determining what time of year to be wary.

In table 7, some of the listings are general, while others are more specific. For example, ash is a general term encompassing such varieties as blue ash, white ash, biltmore ash, as well as the ashleaf maple which is also known as the box-elder. Box-elder has a listing of its own, just as ash, because individuals may relate to them separately. Other examples: There are many kinds of poplar trees, including the common cottonwood, and the pecan tree is actually a member of the hickories.

As far as table 8 is concerned, one should use it only for a general guideline. Shifting weather patterns, population growth, land development, and other factors can alter the starting and ending dates. Also, the pollen count can vary from neighborhood to neighborhood, depending on the kind and quantity of plants in the immediate area.

TABLE 7: COMMON PLANTS WHICH ORIGINATE ALLERGY-PRODUCING POLLEN

Acacia (T)
Adler (T)
Alfalfa (W)
Ash (T)
Bahia (G)
Beech (T)
Bermuda (G)
Birch (T)
Bluegrass (G)
Box-elder (T)
Brome (G)
Burning bush (W) or (S)
Careless weed (W)
Cocklebur (W)
Common reed (G)
Cottonwood (T)
Cultivated oats (G)
Cultivated wheat (G)
Dock (W)
Elm (T)
English plantain (W)

Eucalyptus (T)
Firebush (W)
Goldenrod (W)
Hackberry (T)
Hazel (T)
Hemp (W)
Hickory (T)
Hornbeam (T)
Johnson grass (G)
Lamb's-quarters (W)
Maple (T)
Marsh elder (W)
Meadow fescue (G)
Meadow foxtail (G)
Mexican firewood (W)
Mountain cedar (T)
Mulberry (T)
Nettle (W)
Oak (T)
Olive (T)
Orchard grass (G)

Osage orange (T) or (S)
Ox-eye daisy (W)
Pigweed (W)
Pecan (T)
Pepper tree (T) or (S)
Poplar (T)
Privat (T) or (S)
Ragweed (W)
Redtop (G)
Russian thistle (W)
Ryegrass (G)
Sagebrush (W) or (S)
Sheep sorrel (W)
Spiny amaranth (W)
Sweetgum (T)
Sweet vernal (G)
Sycamore (T)
Timothy (G)
Walnut (T)
White pine (T)
Willow (T)

(G) Grass
(S) Shrub (Some tree or weed varieties can also be classified as shrubbery.)
(T) Tree
(W) Weed

TABLE 8: POLLEN TIME TABLE

CITY	GRASS	TREES	WEEDS*
Atlanta	May-med Sept.	mid Jan.-mid May	mid Aug.-mid Oct.
Baltimore	early May-early July	March-mid May	mid Aug.-Sept.
Boise	May-early Aug.	mid March-May	early July-Sept.
Boston	mid May-mid July	April-May	mid Aug.-late Sept.
Casper, WY	early June-July	late March-April	early June-mid Oct.
Charleston, SC	mid May-early Aug.	mid Feb.-late May	late Aug.-mid Oct.
Chicago	mid May-mid July	mid March-mid June	mid Aug.-Sept.
Cleveland	June-late July	mid March-mid June	early Aug.-Sept.
Dallas	April-Sept.	late Dec.-April & mid Aug.-Sept	Sept.-Oct.
Denver	mid May-late July	mid March-mid May	early June-mid Oct.
Des Moines	late May-Aug.	mid March-mid May	mid Aug.-Sept.
Detroit	mid May-mid July	mid March-early June	mid Aug.-late Sept.
Fargo, ND	June-July	April-May	July-late Sept.
Hartford	mid May-mid July	mid March-late May	early Aug.-Oct.
Indianapolis	mid May-mid July	mid March-mid June	mid Aug.-Sept.
Jackson	early May-Sept.	Feb.-April	Sept.-mid Oct.
Kansas City	mid May-early July	March-May	early July-Sept.

City			
Little Rock	mid May-Sept.	early Feb.-early May	late Aug.-mid Oct.
Los Angeles	March-Nov.	early Jan.-Feb.	July-Oct.
Louisville	mid May-July	March-May	mid Aug.-Oct.
Madison, WS	June-late July	April-May	mid Aug.-Sept
Miami	March-May & Aug.-Dec.	Feb.-March	late May-late Sept.
Minneapolis	late May-late July	April-May	mid June-early Oct.
Montgomery	April-Sept.	mid Jan.-early June	Sept.-early Oct.
Nashville	May-late Aug.	mid Feb.-late May & mid Sept.-early Oct.	mid Aug.-mid Oct.
New Orleans	April-Nov.	Jan.-March	mid July-late Oct.
New York	mid May-mid July	early March-May	mid Aug.-Sept.
Okla. City	May-Sept.	mid Feb.-mid June	July-mid Oct.
Omaha	mid May-mid July	March-May	mid July-Sept.
Philadelphia	early May-mid July	mid March-mid May	mid Aug.-early Oct.
Phoenix	April-Oct.	Feb.-April	March-late Dec.
Pittsburgh	early May-mid July	mid March-mid May	mid Aug.-early Oct.
Portland	late April-Aug.	mid Feb.-April	May-Sept.
Raleigh, NC	mid May-mid July	Feb.-May	early Aug.-Sept.
Reno	June-July	April-May	July-early Oct.
Richmond, VA	mid May-early Aug.	Feb.-mid June	mid Aug.-Sept.
Roswell, NM	mid May-mid Oct.	Feb.-April	mid June-Oct.
Salt Lake City	late April-late July	mid March-late May	mid July-late Oct.

San Francisco	April-Dec.	early Feb.-mid March	May-late Oct.
Seattle	late April-mid Oct.	late Feb.-April	May-mid Oct.
St. Louis	mid May-early July	March-May	early July-Sept.
Tampa	Jan.-Dec.	early Feb.-mid May	early Aug.-Nov.
Washington D. C.	mid May-mid July	early March-late May	early Aug.-Sept.
Wheeling, WV	late May-mid July	mid March-mid June	early Aug.-Sept
Wichita	May-late June	late Feb.-May	mid July-mid Oct.

* For the majority of locations, the dates in this column apply predominantly to ragweed. But there are others significantly implicated in many areas. Keep an eye on the weather section of the newspaper and check with your local health department or chamber of commerce for specifics in regard to not only weeds, but grasses and trees.

Source: Adapted from Rapp, Doris J., Dorothy Bamberg. The Impossible Child: A Guide for Caring Teachers and Parents. (Buffalo, New York. Practical Allergy Research Foundation, 1986.)

APPENDIX G

HELPFUL ADDRESSES

For your convenience, below is a list of helpful organizations and their addresses.

American Academy of Environmental Medicine
P. O. Box 16106
Denver, CO 80216

The above association publishes The Clinical Ecologist, *a quarterly medical journal.*

United States Environmental Protection Agency (EPA)
401 M Street, SW
Suite 200, North East Mall
Washington, DC 20460

The above, a federal government environmental service, is a valuable source of free information.

United States Environmental Protection Agency (EPA)
Office of Radiation Programs
Washington, DC 20460

Write this address for an updated listing of companies specializing in radon testing.

Human Ecology Action League (HEAL)
P. O. Box 49126
Atlanta, GA 30359-1126

This organization supports the programs of those interested in environmental health. They produce a bi-monthly publication entitled The Human Ecologist.

Human Ecology Research Foundation
720 N. Michigan Avenue
Chicago, IL 60611

This organization supports medical research in the field of clinical ecology. Write for a list of publications concerned with the relationship between chronic physical and mental illness and the environment.

Human Ecology Research Foundation of the Southwest
8345 Walnut Hill Lane, Ste. 205
Dallas, TX 75231-4262

A nonprofit orgainization which furnishes patient information and support services. In addition, they sell products for those with environmentally-related health problems. Inquire also about their various publications.

Human Ecology Study Group
Metropolitan Chicago Area Chapter of HEAL
5460 N. Marmora
Chicago, IL 60630

A self-help organization for individuals who have developed chemical sensitivites—as well as food allergies.

National Institute for Occupational Safety and Health (NIOSH)
Parklawn Building
5600 Fishers Lane
Rockville, MD 20857

The research department of OSHA.

New England Foundation for Allergic and Environmental
Diseases
3 Brush Street
Norwalk, CT 06850

*A tax-free, nonprofit organization dedicated to the treatment of
environmental illness. Write for information on the many
publications offered.*

Occupational Safety and Health Administration (OSHA)
U. S. Department of Labor
200 Constitution Avenue NW
Washington, DC 20216

*An organization responsible for controlling health hazards in
the workplace.*

Society for Clinical Ecology
1750 Humboldt Street
Denver, CO 80218

*A body of professionals concerned about environmental effects
on health. They conduct annual seminars, both basic and
advanced.*

The following two establishments are special in that they offer a unique product.

Rexair, Inc.
900 Tower Dr., Suite 700
P. O. Box 3610
Troy, MI 48098

The above address is the company's main headquarters. Rexair sells a portable vacuuming system that traps dust by means of a swirling water bath.

C. D. Headen
1301 Summit
Plano, TX 75074

C. D. Headen is the distributor for "Liquid Ring," a motor vehicle oil additive which increases engine performance and reduces automobile emissions.

NOTES

NOTES

Introduction

1. David Rousseau, W. J. Rea, and Jean Enwright, *Your Home, Your Health, and Well-Being*, p. 217-231.

Chapter 1: The Body's Battle for Survival

1. Theron Randolph and Ralph W. Moss, *An Alternative Approach to Allergies*, pp. 51-52.
2. David Rousseau, W. J. Rea, and Jean Enwright, *Your Home, Your Health, and Well-Being*, p. 41.

Chapter 2: Painting the Pollution Picture

1. Alfred V. Zamm and Robert Gannon, *Why Your House May Endanger Your Health*, p. 14.

Chapter 3: The Unnatural Aspects of Natural Gas

1. Samuel S. Epstein, Lester O. Brown, and Carl Pope, *Hazardous Wastes in America*, pp. 3-5.
2. Theron G. Randolph and Ralph W. Moss, *An Alternative Approach to Allergies*, p. 87.
3. Ibid., p. 239; Alfred V. Zamm and Robert Gannon, *Why Your House May Endanger Your Health*, p. 23.
4. Ibid., p. 55.
5. Theron G. Randolph and Ralph W. Moss, *An Alternative Approach to Allergies*, p. 86.

Chapter 4: The Seriousness of Smoking

1. Dana Miller, "Chronic Epstein Barr Virus—What is it?" *The Human Ecologist*, No. 33 (1986), p. 7. Also note table on p. 9.

2. *Wall Street Journal*, "Smokers May Be at Risk for Osteoporosis," January 23, 1989.

3. Ruth Winter, *Scientific Case Against Smoking*, pp. 37-38; Neil Solomon, *Stop Smoking, Lose Weight*, p.22.

Chapter 5: The Dust Dilemma

1. Theron G. Randolph and Ralph W. Moss, *An Alternative Approach to Allergies*, p. 88. Dr. Randolph refers to this as "fried dust."

2. Alfred V. Zamm and Robert Gannon, *Why Your House May Endanger Your Health*, p. 30.

3. See "ARN Tests 'Low Allergy' Vacuum Cleaners: Ordinary Vacuums Make Your Allergies Worse," *Rodale's Allergy Relief*, vol. 2, no. 11, (November, 1987), pp. 1 and 3-7 for a detailed evaluation of various models.

4. See Sharon Faelten and Editors of Prevention Magazine, *The Allergy Self-Help Book*, pp. 125-128 for an excellent discussion on the subject.

Chapter 6: The Chemical Crisis

1. Theron G. Randolph and Ralph W. Moss, *An Alternative Approach to Allergies*, pp. 98-99.

2. Ellen J. Greenfield, *House Dangerous*, p. 19.

3. Ibid., p. 18; Theron G. Randolph and Ralph W. Moss, *An Alternative Approach to Allergies*, p. 97.

4. Susan Pitman, "Pesticide Update," *The Human Ecologist*, no. 34 (spring 1987): p. 31.

5. Norma Miller, "Environmental Health Tips for Children," *The Human Ecologist*, no. 33 (1986): p. 17.

6. Susan Pitman, "Report on the Fourth National Pesticide Forum," *The Human Ecologist*, no. 32 (1986): p. 26.

7. Norma Miller, "Environmental Health Tips for Children," *The Human Ecologist*, no. 33 (1986): p. 17.

8. Tom Conry, *Consumer's Guide to Cosmetics*, p. 265.

9. Ibid., p. 300.

10. Ibid., p. 117.

11. David Rousseau, W. J. Rea, and Jean Enwright, *Your Home, Your Health, and Well-Being*, pp. 184-186. Table 22-2 offers an excellent breakdown of paints.

12. Natalie Golos, Frances Golos Golbitz, and Frances Spatz Leighton, *Coping With Your Allergies*, p. 177.

Chapter 7: The Mold Menace

1. David Rousseau, W. J. Rea, and Jean Enwright, *Your Home, Your Health, and Well-Being*, p. 41.

2. Ibid., p. 270.

3. Jack Joseph Challem and Renate Lewin, "Winning the Fight Against Multiple Sclerosis," *Let's Live*, (January 1984): pp. 26-28.

Chapter 8: Construction Materials that Can Demolish Health

1. David Rousseau, W. J. Rea, and Jean Enwright, *Your Home, Your Health, and Well-Being*, p. 7.

2. Ellen J. Greenfield, *House Dangerous*, p. 148.

3. Edward J. Bergin and Ronald E. Grandon, *How to Survive Your Toxic Environment*, p. 19 and 22.

4. David Rousseau, W. J. Rea, and Jean Enwright, *Your Home, Your Health, and Well-Being*, p. 68.

5. Melvin Berger, *Hazardous Substances: a Reference*, p. 29.

Chapter 9: The Thing About Plastics

1. Robert S. Swanson, *Plastics Technology*, p. 20.

2. "Outgassing: Aerospace Findings Could Improve Your Health—or Housekeeping," *The Human Ecologist*, no. 5 (October, 1979): p. 3.

3. Ellen J. Greenfield, *House Dangerous*, p. 155.

4. Natalie Golos, Frances Golos Golbitz, and Frances Spatz Leighton, *Coping With Your Allergies*, pp. 37-38.

Chapter 10: Radon Remedies

1. Michael Lafavore, *Radon: The Invisible Threat*, p. 22.
2. Ibid., See chart p.87.
3. "Radon: An Environmental Concern," *The Human Ecologist*, no. 31 (winter 1985-86): p. 14.
4. Michael Lafavore, *Radon: The Invisible Threat*, pp. 166-181. These pages offer a detailed description of the sub-slab ventilation system as well as other similar techniques.

Chapter 12: Guidelines Concerning Gasoline

1. Theron G. Randolph and Ralph W. Moss, *An Alternative Approach to Allergies*, p. 104.
3. Ibid.
4. Ibid., pp. 105 and 228.
5. Ibid., p. 244.
6. Ellen J. Greenfield, *House Dangerous*, p. 45.
7. Alfred V. Zamm and Robert Gannon, *Why Your House May Endanger Your Health*, p. 123.

Chapter 13: Hair, Dander, and Feather Woes

1. Coleman Harris and Norman Shure, *All About Allergy*, p. 68.
2. Ruth Winter, *The People's Handbook of Allergies*, p. 37.
3. Ibid., p. 47.
4. Coleman Harris and Norman Shure, *All About Allergy*, p. 70.

Chapter 14: What About Bacteria?

1. James Howard Otto and Albert Towle, *Modern Biology*, p. 226.
2. Ellen J. Greenfield, *House Dangerous*, p. 151.

Chapter 15: Other Facts Not to be Ignored

1. Ruth Winter, *The People's Handbook of Allergies*, p. 121.

2. David Rousseau, W. J. Rea, and Jean Enwright, *Your Home, Your Health, and Well-Being*, p. 132.

3. Theron G. Randolph and Ralph W. Moss, *An Alternative Approach to Allergies*, p. 94.

4. Sharon Faelten and the Editors of Prevention Magazine, *The Allergy Self-Help Book*, p. 102.

5. Natalie Golos, Frances Golos Golbitz, and Frances Spatz Leighton, *Coping With Your Allergies*, p. 171; Melvin Berger, *Hazardous Substances: A Reference*, p. 30.

Chapter 16: Considering the Future

1. Jon Van, "Turning Potatoes into Plastic," *The Dallas Morning News*, October 30, 1989.

GLOSSARY

GLOSSARY

Acetone: A colorless, flammable liquid used as a solvent; inhalation can cause headaches, fatigue, and bronchial irritation

Air-to-air heat exchanger (or heat-recovery ventilator): Device in which stale air and fresh outside air are exchanged by passing through ducts with a common wall thin enough to readily transfer heat

Alpha particle: A particle of energy which is emitted from certain radioactive substances, including radon

Alpha track test: A radon testing kit which contains photographic paper to detect alpha particles

Alveoli: Generally, small cavities or hollows; in the lung, air cells.

Asbestos: A heat-resistant mineral, which has been implicated in incidents of lung cancer and other diseases when the long, thread-like fibers of which it is composed are inhaled.

Asbestosis: A respiratory ailment caused by the inhalation of asbestos fibers; symptoms include a dry cough and difficulty in breathing

Backdraft: A condition whereby the normal direction of airflow through a chimney (or vent pipe) is reversed because a of change in air pressure

Baseboard duct system: A baseboard of sheet metal installed along a French drain designed to divert radon to the outside

Benzene (or benzol): A colorless, flammable liquid obtained from coal tar and used in making lacquers, dyes, varnishes, etc.; also found in cigarette smoke and implicated in leukemia

Benzo-a-pyrene: A yellow, crystalline hydrocarbon found in coal tar; a constituent of cigarette smoke

Buerger's Disease (Thromboangiitis obliterans): An ailment, rare in nonsmokers, which results in inflammation of peripheral arteries

Cadmium: A blue-white metallic chemical element occurring in zinc ores; suspected of promoting osteoporosis

Carbon monoxide: A colorless, odorless, highly poisonous gas produced by incomplete combustion; most popularly known from its presence in automobile exhaust

Carcinogen: Any substance which induces cancer

Central vacuuming system: A vacuuming device employing a network of ducts throughout a home or other building which sucks dirt into a unit usually located on the outskirts of the premises

Charcoal adsorption test: A radon testing kit which contains activated charcoal granules

Chimney cap: A brick or concrete covering built over the top of a chimney with openings in the side immediately below it; used to prevent backdraft

Cilia: Tiny, hair-like fibers which help eliminate foreign matter from the human system by synchronous movements

Clinical Ecology: A branch of medical science that concentrates on curing disease by improving the environment; also known as environmental medicine

Cresol: A colorless, oily liquid or solid derived from coal tar; can affect the liver, kidneys, spleen, pancreas, and central nervous system

Daughters: See progeny

Drain-tile suction: A method of eliminating radon build-up by connecting drain-tiles to an exhaust fan

Dust mites (dermatophagoides): Microscopic arthropods found in dust

Electronic (electrostatic) air cleaner: An air cleaning device which electrically charges particles, pulls then out of the air, and deposits them on a collector plate

Fibrosis: An abnormal increase in the amount of fibrous tissue, which can be caused by the inhalation of asbestos

Formaldehyde: A colorless, pungent, highly irritating chemical widely employed in manufacturing; a product of combustion found in auto exhaust, cigarette smoke, etc.

French drain: A gap between the floor and the walls of a basement for purposes of drainage, and a potential source of entry for radon; can also be used to exhaust radon by enclosing it in sheet metal and venting it to the outside

Heat pump: A motor-driven device which removes air at one point and deposits it at another; used for heating homes

HEPA (High Efficiency Particulate Air) filter: A high efficiency air filter made of extremely thin glass fibers capable of screening approximately 99.97% of the particulate matter in the air at the 0.3 micron level

HEPA-type filter: A high efficiency air filter less effective than a HEPA filter, but superior to other air purification devices; 95% efficient at the 3 micron level

Humidifier Lung: An ailment, probably caused by the inhalation of the actinomycetes bacteria, which swells the tissue of the lung, causing fever, chills, coughing, and shortness of breath

Hydrocarbons: Harmful compounds of hydrogen and carbon which are products of combustion; found in automobile exhaust, burning natural gas, cigarette smoke, etc.

Hydrogen Cyanide (also known as formonitrile): A highly volatile, colorless, poisonous liquid which boils at 75° Fahrenheit; one of the constituents of cigarette smoke

Kapok: Fibers from the silk-cotton tree used for stuffing pillows, mattresses, etc.

Keratin: The substance in which hair and nails are chiefly composed

Legionnaire's disease (Legionellosis): A lung ailment resembling pneumonia, caused by the inhalation of the bacteria, *Legionella pneumophila*

Mainstream smoke: What some refer to as the smoke emitted from the filter end of a cigarette

Methane: A colorless, odorless, flammable substance which is created by the decomposition of vegetable matter; the major constituent of natural gas

Methylene chloride (dichloromethane): A colorless, volatile liquid used as a solvent; is carcinogenic and has been known to cause birth defects and fetal death

Micron: One millionth of a meter; used in the measurement of microscopic objects

Negative-ion generator: A device which cleans the air by negatively charging particles and pulling them back to a filter

Nitrogen oxides: Products of combustion created when nitrogen is heated to high temperatures; eye, nose, throat, and lung irritant

Olfactory adaptation: A characteristic of the body whereby the sense of smell becomes adapted to an odor to which it is exposed for more than a few minutes, and therefore no longer detects its presence

Outgassing: The gradual release of chemicals in plastics, rubber, synthetic fabrics, paints, glues, etc. due to curing or aging

Ozone: A form of oxygen renown for its existence in the strato-sphere where it shields the earth from excessive ultraviolet radiation; in the lower atmosphere, a pollutant capable of causing such problems as eye irritation, headaches, and respiratory distress

Pentachlorophenol (or PCP): A toxic chemical used to preserve wood; suspected carcinogen and capable of impairing lungs, kidneys, and the liver

Pico Curies per liter (pCi/L): The unit used for measuring the quantity of radon gas in the air; a Curie = 37 billion radioactive decays per second, a pico Curie = 1 trillionth of a Curie, a pico Curie of air = approximately 2 radon atoms per minute decaying in every quart of air

Pipe-in-wall method: A method for eliminating indoor radon concentrations which employs a system of piping installed in basement cinder block walls

Pollinosis: Hay fever

Polychlorinated Biphenyls (or PCBs): Toxic mixtures of organic compounds prepared by the reaction of chlorine with biphenyl

Polyvinyl chloride (PVC): A plastic which is extremely toxic because of its liberal outgassing when new

Progeny: In biology, the products emitted by the radioactive decay of an element; for radon, these products are chiefly bis-muth, lead, and polonium

Propellant: A gas (generally propane, butane, or isobutane) compressed into its liquid state and included in aerosol products as a means to dispatch the contents

Psittacosis (ornithosis): A virus which originates from micro-scopic particles of the feces and nasal secretions of birds so infected; more commonly known as parrot fever

Q fever: A disease characterized by fever, chills, headaches, muscle pain, weakness, loss of appetite, disorientation, and excessive sweating

Rickettsia: A genus of microorganisms comparable in size to that of the virus

Seasonal Affective Disorder (or SAD): A depressed, lethargic feeling some people experience during the winter

Sebum: A semiliquid substance secreted by the sebaceous glands

Second-hand smoke: Exhaled cigarette smoke

Sidestream smoke: What some call the smoke that is emitted from the burning end of a cigarette

Sub-slab ventilation system: A system consisting of a fan and a network of pipes that are used to pull contaminated air out from under a basement floor and vent it outside

Sulfur dioxide: A heavy, colorless, suffocating gas created especially by the burning of coal

Sump: A hole in a basement floor, sometimes accompanied by a pump, which is used for water drainage

Termite shield: Sheet metal covering the wood near a foundation designed to prevent the invasion of termites

Toluene: A liquid hydrocarbon contained in cleaning chemicals, dyes, nail polish, etc.; is a mucous membrane irritant and can affect the central nervous system

Tung oil (or wood oil): A yellow, pungent oil derived from the seeds of the tung tree and used in many products, including construction materials, permanent press garments, and cigarettes; contains toxic esters

Uranium tailings: The sand-like waste produced from uranium mining; the result of chemical separation of the uranium from the ore

Vapor barrier: A material, usually heavy paper, which accompanies insulation for the purpose of preventing moisture from accumulating in wall cavities

Water-trap vacuum cleaner: A portable vacuuming unit which employs a basin of water to remove particulate matter from the air

Working level: Originally devised for uranium miners, the unit of measurement for radon progeny; one working level = 200 pCi/L

BIBLIOGRAPHY

BIBLIOGRAPHY

Aslett, Don, Laura Aslett Simons. *Make Your House Do the Housework*. Cincinnati, Ohio: Writer's Digest Books, 1986.

Berger, Melvin. *Hazardous Substances: A Reference*. Hillside, New Jersey: Enslow Publishers, Inc., 1986.

Bergin, Edward J., Ronald E. Grandon. *How to Survive Your Toxic Environment*. New York: Avon Books, 1984.

Blum, Alan. *The Cigarette Underworld: A Front Line Report on the War Against Your Lungs*. Secaucus, New Jersey: Lyle Stuart, Inc., 1985.

Brenner, David J., *Radon: Risk and Remedy*. New York: W. H. Freeman and Company, 1989.

Brumbaugh, James E. *Heating, Ventilating, and Air Conditioning Library, Volume III*. Indianapolis, Indiana: Bobbs-Merrill Company, Inc., 1976.

Christensen, Clyde M. *The Molds and Man*. Minneapolis, Minnesota: University of Minnesota Press, 1965.

Clifton, Charles. *Introduction to Bacteria*. New York: McGraw-Hill Book Company, 1958.

Cohen, Bernard. *Radon: A Home Owners Guide to Detection and Control*. Mt. Vernon, New York: Consumers Union of United States, Inc., 1987.

Conry, Tom. *Consumer's Guide to Cosmetics*. Garden City, New York: Doubleday and Company, Inc., 1980.

Dangerfield, Stanley, Elsworth Howell. *The International Encyclopedia of Dogs*. New York: Howell Book House, 1981.

Epstein, Samuel S., Lester O. Brown, Carl Pope. *Hazardous Waste in America*. San Francisco, California: Sierra Club Books, 1982.

Erb, Russell C. *The Common Scents of Smell*. Cleveland, Ohio: World Publishing Company, 1968.

Faelten, Sharon, Editors of Prevention Magazine. *The Allergy Self-Help book*. Emmaus, Pennsylvania: Rodale Press, 1983.

Golos, Natalie, Frances Golos Golbitz, Frances Spatz Leighton. *Coping with Your Allergies*. New York: Simon and Schuster, 1979.

Greenfield, Ellen J. *House Dangerous*. Mt. Vernon, New York: Consumers Union of United States, Inc., 1987.

Harris, Coleman M., Norman Shure. *All About Allergy*. Englewood Cliffs, New Jersey: Prentice-Hall

Jaquith, Elaine Bonavita. *Allergic to the 20th Century*. Hurst, Texas: ABC Printing, 1986.

Lafavore, Michael. *Radon: The Invisible Threat*. Emmaus, Pennsylvania: Rodale Press, 1987.

Makower, Joel. *Office Hazards: How Your Job Can Make You Sick*. Washington D. C.: Tilden Press, 1981.

Mandell, Marshall, Lynne Waller Scanlon. *Dr. Mandell's 5-day Allergy Relief System*. New York: Simon and Schuster, 1979.

Moore, Alma Chestnut. *How to Clean Everything*. Forge Valley, Massachusetts: Murray Printing Company, 1968.

Ochsner, Alton. *Smoking: Your Choice Between Life and Death*. New York: Simon and Schuster, 1970.

Otto, James Howard and Albert Towle, *Modern Biology*. New York: Holt, Rinehart and Winston, Inc., 1969.

Pfeiffer, Guy O., et al. *Household Environment and Chronic Illness*. Springfield, Illinois: Charles C. Thomas, 1980.

Randolph, Theron G., Ralph W. Moss. *An Alternative Approach to Allergies*. New York: Bantam Books, 1980.

Rapaport, Howard G., Shirley Motter Linde. *The Complete Allergy Guide*. New York: Simon and Schuster, 1970.

Rapp, Doris J., Dorothy Bamberg. *The Impossible Child: A Guide for Caring Teachers and Parents*. Buffalo, New York: Practical Allergy Research Foundation, 1986.

Richman, Beth, Susan Hassol. *Everyday Chemicals*. Snowmass, Colorado: The Windstar Foundation, 1989.

Rousseau, David, W. J. Rea, Jean Enwright. *Your Home, Your Health, and Well Being*. Vancouver, B. C.: Hartley and Marks, Ltd., 1987.

Simonds, Herbert R. *Concise Guide to Plastics*. New York: Reinhold Publishing Co., 1963.

Solomon, Neil. *Stop Smoking, Lose Weight*. Toronto, Ontario: Academic Press, 1981.

Stabile, Toni. *Everything You Want to Know About Cosmetics*. New York: Dodd, Mead and Company, Inc., 1984.

Stellman, Jeanne, Mary Sue Henifin. *Office Work Can be Dangerous to Your Health*. New York: Pantheon Books, 1983.

Swanson, Robert S. *Plastics Technology*. Bloomington, Illinois: McKnight and McKnight Publishing, 1965.

Turner, Roger Newman. *Hay Fever Handbook*. Wellingborough, Northhamptonshire: Thorson's Publishing Group, 1988.

Wallace, Bruce. *People, Their Needs, Environment, Ecology*. Vol. 1. Englewood Cliffs, New Jersey: Prentice-Hall, 1972.

Whitney, Leon F. *The Complete Book of Cat Care*. Garden City, New York: Doubleday and Company, Inc., 1980.

Winter, Ruth. *The People's Handbook of Allergies and Allergens*. Chicago, Illinois: Contemporary Books, Inc., 1984.

——— . *Scientific Case Against Smoking*. New York: Crown Publishers, Inc., 1980.

——— . *So You Have Sinus Trouble*. New York: Grosset and Dunlap, 1973.

Zamm, Alfred V., Robert Gannon. *Why Your House May Endanger Your Health*. New York: Simon and Schuster, 1980.

Other Source Materials

Air Pollution in Your Home? American Lung Association Pamphlet. No. 1001, November, 1984.

Allergies Alleged to be Cause of Psychoses. Medical World News. January 30, 1970.

Allergy a Factor in Cerebral Palsy? Prevention Magazine. Vol. 34, No. 2, February, 1982.

An Introduction to Ecologic Mental Illness: Demonstrable Cerebral Reactions from Foods and Chemical Exposures--a Motion Picture Documentation. Review of Allergy. Vol. 23, no. 11, November, 1969.

ARN Tests "Low Allergy" Vacuum Cleaners: Ordinary Vacuums Make Your Allergies Worse! Rodale's Allergy Relief. Vol. 2, No. 11, November, 1987.

Celis, William III. *California Orders Deodorant Makers to Cut Propellants*. Wall Street Journal. November 15, 1989.

Challem, Jack Joseph, Renate Lewin. *Residual Polio: One Doctor, One Patient, and a Natural Approach*. Let's Live. November, 1984.

——— . *Winning the Fight Against Multiple Sclerosis*. Let's Live. January, 1984.

——— . *Your Pains May be Misdiagnosed Allergies* (interview with Dr. Marshall Mandell). Let's Live. May, 1979.

Dickey, Lawrence D. *Ecologic Illness*. Rocky Mountain Medical Journal. Vol. 68, no. 10. October, 1971.

Environment: Iowa Ranked No. 1 in Radon Survey. Science News. Vol. 136, no. 20. November 11, 1989.

Environment: Keep Risky Rocks Under Wraps. Science News. Vol. 136, no. 20. November 11, 1989.

Grossmann, John. *The House that Glowed*. Health Magazine. October, 1988.

Henig, Robin Marantz. *The Enemy Within*. AARP Bulletin. Vol.31, no. 8, September, 1990.

Hoffman, Joseph H., Andrew E. Filderman. *Lung Cancer Update on Risk and Prevention.* Contemporary Internal Medicine. January, 1990.

How to Fix a Wet Basement. Consumer Reports. February, 1990.

Indoor Air Worse Than Outdoor: New CPSC Study Report. The Human Ecologist. No. 26. Summer, 1984.

Lead Poisoning Poses Risk to U. S. Children Despite Precautions. Wall Street Journal. August 18, 1988.

Mandell, Marshall. *The Allergy Letters*. Volume 50. November 1987.

——— . *American Academy of Environmental Medicine 18th Annual Seminar (reprint)*. October 14, 1984.

——— . *Cerebral Manifestations of Hypersensitivity to the Chemical Environment: Susceptibility to Indoor and Outdoor Air Pollution*. Review of Allergy, Vol. 22, no. 9. September, 1968.

——— . *Cerebral Reactions in Allergic Patients*. Journal of the International Academy of Metabology, Vol. III, no. 1. March, 1974.

——— . *Clinically Important Species--Specific Mold Allergens in Asthma and Rhinitis, Demonstrated by Provocative Nasal Inhalation Tests: Etiologic Diagnosis of Individual Allergic Manifestations.* Kansas Medical Society. Journal of Kansas Medical Society. October, 1968. Vol. LXIX, no. X.

——— . *Does Environment Make Us Ill?* The New England Foundation of Environmental Diseases, 3 Brush Street, Norwalk, Connecticut. (Printed from a tape recording of the presentation given by Dr. Mandell to the 6th International Water Quality Symposium in Washington, D. C. on April 18, 1972.)

Mandell, Marshall, Anthony A. Conte. *The Role of Allergy*. Journal of the International Academy of Preventive Medicine. July, 1982. Vol. VII, no. 2.

Manges, Michele. *Drive to Eliminate Asbestos from Schools Slowed by Shortages of Funds, Contractors*. Wall Street Journal. August 8, 1988.

Miller, Dana. *Chronic Epstein Barr Virus--What is It?* The Human Ecologist. No. 33, 1986.

Nero, Anthony V., Jr. *Controlling Indoor Air Pollution*. Scientific American. Volume 258, No. 5. May, 1988.

Outgassing: Aerospace Findings Could Improve Your Health-- or Housekeeping. The Human Ecologist. No. 5. October, 1979.

Pesticide Update. The Human Ecologist. No. 33, 1986.

Pitman, Susan. *Pesticide Update*. The Human Ecologist. No. 34.
　　Spring, 1987.

Pitman, Susan. *Report on the Fourth National Pesticide Forum*.
　　The Human Ecologist. No. 32. 1986.

Radon: An Environmental Concern. The Human Ecologist. no.
　　31. Winter, 1985.

Van, Jon. *Turning Potatoes to Plastic: Food Waste Could
　　Reduce Pollution*. The Dallas Morning News. Monday,
　　October 30, 1989.

INDEX

INDEX

Note: bold listings indicate illustrations